THE SHIFTING OF THE POLES:

The Earth actually tips over in space, creating an arctic Florida!

"It is 5 P.M. in Boston and the people rushing home from work hardly seem to notice how the setting sun seems to hang on the horizon. After several hours people start to wonder why darkness is not coming, and they start to fear the faint dull roar they hear . . ."

* * *

"Psychic phenomena are facts. Psychics are not always right. Neither are scientists. Perhaps if both worked together, the results could be more useful than either could obtain alone. This, it seems to me, is the most important message of this fascinating book."
—DAVID M. STEWART, Ph.D. Former Director, MacCarthy Geophysics Laboratory and Seismic Observatory, University of North Carolina.

* * *

WE ARE THE EARTHQUAKE GENERATION

INCLUDING *NEW MATERIAL* ESPECIALLY WRITTEN FOR THIS EDITION:

- WAKE UP, U.S.A.!
- THE TEN MOST DANGEROUS CITIES IN THE U.S.A., PLUS TWO
- HOW TO SURVIVE AN EARTHQUAKE

Berkley Books by Jeffrey Goodman, Ph.D.

PSYCHIC ARCHAEOLOGY
WE ARE THE EARTHQUAKE GENERATION

WE ARE THE EARTHQUAKE GENERATION

WHERE AND WHEN THE CATASTROPHES WILL STRIKE

JEFFREY GOODMAN, Ph.D.

A BERKLEY BOOK
published by
BERKLEY PUBLISHING CORPORATION

FOR MY FATHER AND OUR FATHER

This Berkley book contains the complete
text of the original hardcover edition.
It has been completely reset in a type face
designed for easy reading, and was printed
from new film.

WE ARE THE EARTHQUAKE GENERATION

A Berkley Book/published by arrangement with
Seaview Books

PRINTING HISTORY
Seaview Books edition published 1978
Berkley edition/October 1979
THIRD PRINTING

For information address: Seaview Books,
A Division of Playboy Enterprises, Inc.,
747 Third Avenue, New York, New York 10017.

ISBN: 0-425-04203-0

A BERKLEY BOOK® TM 757,375
Berkley Books are published by Berkley Publishing Corporation,
200 Madison Avenue, New York, New York 10016.
PRINTED IN THE UNITED STATES OF AMERICA

Publisher's note

The information contained in *WE ARE THE EARTHQUAKE GENERATION* is not pleasant to read, but we feel that every individual has a right to know about the dangers of earthquakes, both in present day America and throughout the world. With this in mind, we have asked Dr. Goodman to update this paperback edition, adding any new developments possible before we went to press, and to add new material about the most dangerous cities in the U.S.A. (*see Appendix A*). Another area of concern is what to do in the event an earthquake happens, so we've added a section on how to survive an earthquake *(see Appendix B).*

The possibilities outlined in this book are chilling, but we feel you should be made aware of them.

—Berkley Publishing Corporation

ACKNOWLEDGMENTS

Many thanks to John White for his encouragement, editorial assistance, and total support on all fronts. And thanks to the patient and helpful members of my family. Ms. Jo Anne Jansen typed the manuscript drafts and Sheila Pitt drew most of the illustrations. Substantial help and encouragement have also come from my agent Knox Burger, Bob Christian, Art Gilliam, Charmion McCusick, Pat Reidy, and Jaine Smith. Finally, very special thanks to the many psychics who have shared their insights with me, especially Aron and Doris Abrahamsen, Clarisa Bernhardt, Ray Elkins, Susan Harris, Beverly Jaegers, and Bella Karish.

Earthquake-prone areas include some of the densely populated regions in the world, such as Japan, the western United States, and the shores of the Mediterranean Sea. It has been estimated that over 500 million persons could well suffer damage to their property, while a significant proportion of them are in danger of losing their lives in severe earthquakes.

—*Disaster Preparedness Report to Congress*, Office of Emergency Preparedness, Executive Office of the President, January 1972

Unless there is a gigantic conspiracy involving some thirty university departments all over the world, and several hundred highly respected scientists in various fields, many of them originally hostile to the claims of the physical researchers, the only conclusion the unbiased researcher can come to must be that there does exist a small number of people who obtain knowledge existing either in other people's minds, or in the outer world, by means as yet unknown to science.

—Arthur Koestler, *The Roots of Coincidence*, 1972

He who is conscious of danger creates peace for himself; he who takes things lightly creates his own downfall.

—*I Ching*, or *The Book of Changes*

Contents

Preface

This book presents the views of a group of psychics or clairvoyants who have made predictions about the future of our planet and what that might mean for humanity. It also presents scientific evidence in support of these views. The material deals primarily with geological processes that may lead to vast changes in the geography of the globe, culminating in a cataclysmic pole shift—a major displacement of either the rotational axis or the earth's crust—about 2000 A.D.

My purpose in presenting this information is not to alarm people; nor do I claim inevitability for the psychic projections presented here. Time alone will tell. I can say at the outset, however, that the psychics with whom I worked all have remarkable records of prediction and that these have been subjected to scientific verification. Together their prognostic statements about the earth produce an integrated and coherent picture of its future.

Since most of the major events predicted are still more than twenty years away, we can take advantage of this period to evaluate these psychic warnings of disaster. Undoubtedly, a number of them will prove to be incorrect. Statistically speaking, this is inevitable. But regardless of a certain percentage of inaccuracies and failures, the mere volume of the predictions alone and the information provided by them must act to galvanize the scientific and political communities into taking preventive steps to counter those earth changes that increasingly threaten human lives.

From a scientific point of view, what is important about these predictions is that to some degree—because they are made in relation to certain measurable geological, biological, meterological, astronomical, and biophysical factors—their reliability is potentially testable. And in time, such testing may become sufficiently developed that major refinements of the predictions may be possible. We may find, for example, that the relationships described in the predictions are correct but the time scale is not. Or vice versa. Clearly, only the passage of years will give us the full story.

The psychics maintain, however, that as the story unfolds it will be subject to some degree of human modification. Thus I introduce a new term, "biorelativity," to identify a previously unrecognized interaction—subtle but powerful and all-pervasive—between the human race and the planet. Through biorelativity, the primary energy source and controlling influence of which is humanity, many seemingly natural physical events can theoretically be influenced and even arrested. This is a radical position that will be derided by many scientists. But if the psychics' earth-change predictions begin to prove true in the next few years, these skeptics should be intellectually honest enough to consider the subject of biorelativity without prejudice.

The American psychic Edgar Cayce used to say: "First do the research, then comes enlightenment." This is useful advice to bear in mind as we compare the psychics' view of planetary-cosmic forces and processes with contemporary scientific perspectives. As stated, this psychic vision may be inaccurate to some degree. But their vision does provide a framework for understanding, insight, and even possible enlightenment. It is something we can use to guide our study, something we can add to or change when evidence warrants. If psychics' views provide any help at all, we will at least be moving in the right direction at a time when official earth science is all too puzzled about planetary-cosmic forces and processes. Perhaps these clairvoyants can start us looking more carefully in unnoticed areas so that (1) the set of possible consequences they predict is nullified or at least reduced in severity and probability, and (2) scientific understanding becomes more sophistiated and reflective of reality.

Thomas Kuhn, author of *The Structure of Scientific*

Revolutions, shows in his analysis of the history of science that when normal scientific paradigms offer no solution to important problems, extraordinary research has taken place. This book is an invitation to undertake such extraordinary research. It presents important new information, but many questions remain unanswered. My hope is this: that enough people with social conscience and a true sense of scientific inquiry will recognize the usefulness of what is offered here and undertake a careful investigation. The relatively low dollar cost of evaluations versus the potentially high value of the payoff in terms of lives and property saved should more than compensate for the low probability of success that most scientists expect from psychically guided research.

WAKE UP, U.S.A.!
(*The Score to Date*)

On June 3, 1978, *The New York Times* in a cover story announced that the President's Office of Science and Technology "to minimize damage and loss of life when, as is believed inevitable, a major earthquake strikes the United States" has prepared plans "for creation of new Federal agencies, widespread reinforcement of structures and extensive research on earthquake prediction, control and hazard reduction." Ten metropolitan areas were judged to be in the greatest danger: San Francisco, Los Angeles, Salt Lake-Ogden, Puget Sound (Seattle-Tacoma), Hawaii, St. Louis-Memphis, Anchorage-Fairbanks, Boston, Buffalo, and Charleston (*see Appendix A for details*). Nevertheless, citizens are tending to ignore new danger signals occurring across the entire country and many do not know how to prepare for earthquakes *(see Appendix B).*

Since July 1978, when the book, *WE ARE THE EARTHQUAKE GENERATION*, was published in hardcover, things seem to be heating up, just as the book said they would.

- On April 28, 1978, *Science Magazine* carried a report which concluded that there is a 5-to-11 percent chance that New York City will experience a sizeable earthquake in the next 40 years. The authors of the report, Drs. Yash Aggarwall and Lyne Sykes of Columbia University's Lamont-Doherty Geological Observatory, placed instruments on the nearby Ramapo fault and discovered that it was far from inactive: they measured a number of small earthquakes along the fault. The authors also noted that 3 nuclear power plants at Indian Point, New York, are situated close to this fault. The greater New York City area has already recorded sizeable earthquakes in 1737, 1884, and 1927.

- On September 15, 1978, *Science Magazine* carried a report about the threat of earthquakes in the eastern United States and concluded that the hazards are quite real. The potential for earthquake hazard was particularly noted for the New York-New Jersey area, the Boston-Cape Anne area, Charleston, S.C., and St. Louis, Missouri. Reactivation of old seismic sources by various means was proposed as the underlying cause of eastern seismicity. The article noted recent studies showing the association between the large historical quakes along the New Madrid fault outside St. Louis and recent tremors. And as if to emphasize their point on September 20, 1978, St. Louis experienced a 3.5 earthquake.
- On August 18, 1978, a series of small earthquakes struck Mount Shasta, a 14,000 foot volcano in Northern California that hasn't erupted in 200 years. These were the first recorded earthquakes in the area and they caused large cracks in Mt. Shasta—some nearly a mile long.
- On August 24, 1978, a University of Chicago geophysicist told *The Los Angeles Times* that "volcanoes on the northwest coast of the U.S. could erupt and destroy Seattle and Portland." The scientist found that the large volcanic mountains in the Cascades are far from dead and could erupt violently. He noted that the last violent volcanic eruption on this continent created Crater Lake in Oregon, a scant 6,000 years ago. In January 1979, the United States Geological Survey issued a warning regarding two of these volcanic mountains. Dr. Dwight Campell of the survey speculated that there could be an eruption by the end of this century.
- On September 23, 1978, the director of the U.S.

Geological Survey sent out a letter to California officials warning of a possible new volcanic eruption on Mt. Shasta. The letter was accompanied by a report urging a series of steps to minimize the danger. Dr. C. Dan Miller, the USGS geologist who compiled the report, said that "it could come any day or it could be several hundred years." Ominously, Miller said that the recent earthquakes on Shasta occurred *after* he finished his report warning of the possible "explosive eruption" of Shasta.

- Even scarier is that the psychics I worked with said San Diego would be the first of the California cities to be hit by a major quake. And on August 19, 1978, after a series of small and unusual quakes, San Diego geophysicists have changed their minds about the threat to San Diego and now consider it to be "living on borrowed time" where a quake of at least 7.3 on the Richter scale is considered due in less than 60 years. And as if this wasn't enough for San Diego, on December 2, 1978, the *San Diego Union* announced that "Experts Fear Fault Perils Downtown." Leighton Associates of Irvine, California, the independent geological engineering firm hired to study the downtown San Diego area for an urban redevelopment project, discovered that the Coronado Fault previously believed to end offshore may in fact ("strong possibility") extend through the heart of downtown San Diego and be active. The firm also noted several other fault traces cutting through the downtown area.

- And geologists still can't explain the mysterious landslides that destroyed over 30 homes and did four million dollars in damage in Laguna Beach, California, on October 3, 1978. Just 10 days after, geologists checked the area pursuant to the issuance of new building permits, "the earth shifted as much as 40 feet vertically and 30 to 40 feet laterally. One fissure was 50 feet long and another crack was 30 feet long and 30 feet deep."

- And on September 25, Dr. Kerry Sieh, a Cal Tech geologist, put an end to speculation about how often southern California suffers major devastating earthquakes. Sieh cut a deep trench in a marsh that straddles the San Andreas Fault just 55 miles from Los Angeles. Distinct breaks in the deposits corresponded to earthquake shifts. After these breaks were dated, Sieh concluded that major quakes occurred on an average interval of 167 years and sometimes as short as 55 years. Because the last big quake hit southern California in 1857 (Fort Tejon), Sieh concludes "a major earthquake can be expected there [southern California] within the next 50 years." Other geologists note how Sieh may well be right since his telltale marsh is only 15 miles from Palmdale, the center of a region that has been bulging for the last 17 years in what may be a prelude to just such a major earthquake.
- On December 1, 1978, Mexico experienced its worst quake in two decades. Two buildings in downtown Mexico City swayed so much they collided.
- A recent study by Mitre Corp., a private research organization, says that 181 of the 594 booming noises heard off the East Coast between December 1977 and June 1978 might be connected with sudden movements of the earth's crust under the ocean. The study noted hundreds of documented accounts from around the world of booming noises being associated with earthquakes.
- Besides, the earth "heating up," I also warned of man's activities in bringing about more earthquakes via activities such as building dams, drilling oil wells, and underground nuclear testing. Thus, what are we to think when on August 13, 1978, Santa Barbara had an offshore earthquake which did almost $15 million in damage and on August 18, 1978, the U.S. Geological Survey announced that it is possible that this quake was triggered by loading mud in the drill hole of a nearby

offshore exploratory oil well that threatened to get out of control just four days before the quake. Hence, the discovery of oil and gas off Long Island, New York, and New Jersey could lead to substantial drilling activity and be a cause of great concern.

- And what are we to think of proposed nuclear power plants for the country in quake prone areas when in 1961 a Russian nuclear power plant blew up in the Ural Mountains and leveled an entire town where it seems that this explosion was triggered by an earthquake. Even if it didn't cause an explosion, an earthquake could damage the buildings of a nuclear facility and unleash radioactive fallout. Interestingly, in January of 1979, the Nuclear Regulatory Commission withdrew its endorsement of the main U.S. nuclear reactor safety study because it "may understate the risk of atomic power plant disasters."
- On March 13, 1979, the Nuclear Regulatory Commission ordered five atomic plants to immediately shut down because their cooling systems may be unable to withstand earthquakes. One plant was in Maine, and residents laughed until a moderate-sized earthquake struck the area a few weeks later.

- And who do the relatives of the 25,000 Iranians who lost their lives in the September 16, 1978, 7.5 earthquake outside the city of Tabas complain to, considering that Dr. Heinz Kaminski, a German seismologist, and seismologists from the United Kingdom believe that this quake was triggered by an underground nuclear test which took place 36 hours before the quake 1,500 miles away in southern Siberia?

I hope to inform the public and dislodge officials from their state of apathy which recently caused them to ignore taking the special precautions urged in the U.S.G.S. report warning about a potential new volcanic eruption at Mt. Shasta. Or recently supported a proposal to build what would be the tallest self-supported building in the world, "Space Tower 2000" in Long Beach, California, in an area to be partially created with landfill along the harbor's edge. Imagine a 2000-foot high building with a "super deluxe hotel" at the top in the same Long Beach that suffered a devastating earthquake in 1933.

—JEFFREY GOODMAN, PH.D.

CHAPTER 1

A Quake by Any Other Name

"It felt like the end of the world."

"My life *did* pass before my eyes."

"There was a thunderously loud rumble of buckling earth and masonry."

"It hurts your ears, teeth, and bones."

"It felt like I was standing on jelly."

"The building was moving."

At first I thought it was a truck passing. But then the roof caved in on my mother. She died instantly."

"There were three to four minutes of continuously intensifying quakes."

"An atmospheric shock wave was recorded more than two thousand miles away and water in wells as far away as South Africa jumped abruptly."

"I saw dust rise like a blanket being lifted all across the city . . . The dust rose to about a thousand feet until I saw nothing but dust and fires."

"Within minutes the streets were overflowing with people."

"Thousands roamed the streets as if dazed, as smoke billowed from the rubble."

"I saw buildings collapse and fold down like accordions."

"Fires were widespread."

"Almost all the buildings in the center of the downtown sections are destroyed."

"The nine-story hotel we were in literally split down the middle."

"There are bodies in almost all the debris."

"It's not over here. Almost hourly, the earth does its dance—sometimes an easy sway, sometimes a sharp rumble. Buildings bow, raining fresh bricks, glass, and plaster into the littered streets."

"Women shriek. They clutch their rosaries and weep. Men jump for the center of the street, reciting prayers aloud. Adults, officials, doctors, soldiers, police, all know the drill now: Move quick, shout something to heaven, check the family."

"The resulting tidal waves were far more awesome and terrifying than the quake. There were roaring walls of water."

"They're eating rats and anything they can get their hands on."

"Water contaminated by ruptured sewage lines was the chief immediate threat. We feared a cholera epidemic."

"It is a miracle we are still alive."

These are actual comments taken from newspaper accounts from people who were caught in earthquakes during recent years. But many who were caught in earthquakes didn't survive to report their experiences. The death toll from history's great quakes is staggering. In 1556, an earthquake in Shensi, China, left almost one million dead. The 1737 Calcutta quake took three hundred thousand lives; the 1755 Lisbon quake killed sixty thousand; the 1883 Dutch Indies quake, thirty-six thousand. Look at the figures for just a short listing of some big earthquakes and volcanic eruptions this century.

Year	*Location*	*Approximate Death Toll*
1902	Martinique	40,000
1905	India	20,000
1906	San Francisco	700
1908	Sicily	75,000
1915	Italy	30,000
1920	Kansu, China	180,000
1923	Tokyo	143,000

1932	Kansu, China	72,000
1935	India	60,000
1939	Turkey	30,000
1939	Chile	30,000
1950	India	30,000
1960	Agadir, Morocco	12,000
1964	Alaska	114
1968	Iran	12,000
1970	Peru	67,000
1972	Managua, Nicaragua	10,000
1975	Mukden, China	Unconfirmed reports of many thousands

The year in which this book was begun—1976—according to the United States Geological Survey (USGS), saw earthquake activity of greater proportions than in the past half century. Likewise, the death toll for 1976 was staggeringly high:

Location	*Approximate Death Toll*
Turkey	5,000
U.S.S.R.	Unconfirmed reports of many thousands
Guatemala	23,000
New Guinea	9,000
Tangshan, China	750,000
Italy	900
Bali	6,000
Philippines	3,000

No wonder that throughout history the human race—and all other animal species—has been terrified by earthquakes. The loss of life and property easily equals that lost in major warfare. For example, in the 1906 San

Francisco quake more than $500 million in damage was done. The occurrence of such a quake today would cause damage in the billions.

The prediction of earthquakes is obviously of major importance to all humanity. But the events of the last few years give us reason to be doubly concerned. Consider these facts:

1976 showed such a dramatic increase in earthquake activity that some earth scientists have publicly stated that the world is entering a period of increased seismic disturbance.

Some major volcanoes are beginning to show signs of renewed activity. In the opinion of some scientists, global volcanic activity has been on the increase since 1955.

Major earthquakes are occurring in places they have rarely occurred before, such as the 1976 Italian quake.

Chinese and Japanese scientists are independently anticipating many more major quakes in their countries and are giving earthquake prediction a high priority in their research efforts.

The world's climate is suddenly becoming unstable. Over the last few years, record rains and floods have soaked some areas while droughts have parched others. In 1976, for example, Britain experienced the worst drought it had in five hundred years while in some parts of Africa (Mali, Senegal, Mauritania, Upper Volta, and Nigeria) torrential rainfalls spawned an almost biblical plague of rats, locusts, and caterpillars. Some experts even predict that the world will soon enter another ice age where the polar ice caps will dramatically and abruptly increase in size. Others forecast major famines.

The ground is beginning to swell or bulge more and more each day near Palmdale, California, and some geophysicists forecast the occurrence of quakes there bigger than the terrible 1906 San Francisco quake (8.3 on the Richter scale). Dr. Robert M. Hamilton, chief of the USGS's Office of Earthquake Studies, recently said, "Californians should not ignore or underestimate the earthquake threat." He told California to prepare for an "inevitable" great earthquake (8.0–8.9 on the Richter scale) that "will indeed be a disaster."

The densely populated eastern United States is rapidly becoming recognized as a potential quake area as geologists map out deep fault systems. Places such as Delaware have received their first recorded quakes. Even New York City has major fault systems running through it. (Can you imagine the consequences if a quake hit where our proliferating nuclear power plants are located?)

With recent quake activity in Michigan, Tennessee, Arizona, and Colorado, the devastating quakes that struck the Midwest in the 1800s have been brought back to mind. Could the fault zones in those areas, quiet for more than a century, be storing energy for big new earthquakes in the coming years?

A major geomagnetic pole reversal, in which the poles change polarity (from electrical charge + to –), is long overdue, and experts believe that present magnetic field changes are a prelude to such a reversal.

Finally, with the newly recognized association between earthquake activity and the movement of the planets, which involves a tidal-like effect (earth tides), and with the rarest and most severe of such alignments coming up in 1982 and 2000, serious concern is certainly justified.

* * *

A pattern of increasing natural geological disaster seems to have begun. That we may be in for a concentrated period of earthquakes and other cataclysms in the immediate future is suggested by data from diverse scientific disciplines such as geology, geophysics, paleoclimatology, archaeology, astronomy, meteorology, and biology. Not only does information from these fields point in the same destructive direction, but there are also clear connections between these fields. These connections, overlooked till now, indicate that we may be missing the "forest for the trees" because of the fragmentation of scientific perspective resulting from academic specialization. For example, a good deal of geological evidence indicates that our planet has been through global catastrophe several times before, and such evidence correlates with the archaeological account in which the events of one such destructive period—at the close of the last ice age—were actually recorded by ancient man.

There is even evidence that the earth periodically exceeds the equilibrium point of its diurnal wobble and tumbles over in space—a sudden shift that would instantly rearrange the world's oceans, ice caps, and climates. Such an event, if viewed from space, would be stupendous to behold. Think of the earthquakes and volcanic eruptions, the rifts where the earth would split apart, the mountain chains that would be thrust upward, the cyclonic winds, the mountainous tidal waves, and the electrical storms of inconceivable intensity.

Misinformation, misinterpretation of available evidence, and the frequent disregard of data that do not coincide with popular theories may be lulling us into a sense of false security about what our planet holds in store for us. The thesis of this book is that soon—in fact in the next generation—we may face such cataclysmic changes that civilization as we know it would end. Some investigators think that we are already living on borrowed time.

A pattern of increasing disaster leading up to full-scale

planetary catastrophe would come as no surprise, of course, to the ten million people who have bought Hal Lindsey's book *The Late Great Planet Earth* or to others who place faith in the biblical prediction of Armageddon. In fact, the teachings of a number of religious traditions, from the Hopi Indians of Arizona to groups in Africa, foretell such a period, just as their teachings and those of virtually every tribe and race of humanity tell of such a destructive period in the past. The story of Noah and the legend of Atlantis are only two of the most notable.

This pattern of increasing disaster would also come as no surprise to the millions of people who have followed the predictions of psychics such as America's Edgar Cayce and France's Nostradamus. Edgar Cayce (1877–1945), who gained fame for his many accurate medical predictions, already has predicted several major quakes. And Nostradamus, who gained fame as the royal physician to Charles IX of France, accurately prophesied in the sixteenth century such major world events as the Russian Revolution and Hitler's rise to power.

Both Cayce and Nostradamus prophesied that our planet will undergo a period involving major earthquakes and volcanic activity before the year 2000. The mere mention of the word "psychic" offends some people, but if they can sit still long enough to examine the incredibly accurate and scientifically sound perceptions to be found in the comments of some psychics, their opinion might change, just as mine did. The possibility of major earthquakes and cataclysms is too serious a subject to ignore. Even the most skeptical should, in the name of common good, at least consider the psychic view. In our effort to understand and predict earthquakes—and thereby to save human lives—it should not matter if the knowledge comes from an unconventional source.

Interestingly, psychics today around the world are also describing in varying ways a time of increasing disaster for the earth, which they say will reach its zenith in the year 2000. Unlike scientists, psychics have been vocal in predicting an earth-change generation and specific in their description of events to come. These psychics, whose records in other kinds of psychic prediction are impressive, are united in their general presentation of a nearing cataclysmic period. Dare we dismiss such a strong and reiterated theme as so much nonsense?

Since psychics tend to read what other psychics say in print, we must look for independently derived predictions. Therefore, all the more recent psychic predictions about coming earth changes must be measured against the base line established by Edgar Cayce and Nostradamus.

At the moment, we do not fully understand the nature of earthquakes. We do know that more than a million quakes shake the world each year, although only one hundred thousand are strong enough to be felt and fewer than one thousand cause any damage. Each year averages one "great" (8.0–8.9) earthquake and a dozen or more lesser quakes that may be termed "major" (7.0–7.9). Eighty percent of the earthquakes occur within the area bordering the Pacific Ocean dramatically named "the Ring of Fire." And more than 90 percent of all Americans live in an area where there is a significant danger of earthquakes.

Seismologists know which areas of the globe are most prone to earthquakes. But they are unable to predict when a tremor will actually strike because they don't fully understand them. At best, all that seismologists, geologists, and geophysicists can do is measure the physical factors associated with earthquakes and hope for a clue. In fact we can't even be sure if scientists truly understand the nature of any of the most important geological happenings, many of which are intimately related to quake activity. We must also confront the fact of "geofantasy."

What is geofantasy? After I received my degree in geological engineering from the Colorado School of Mines and became involved in oil exploration, I quickly learned about it and gave it the name. Many geological explanations, I discovered, had reality only in the minds of the "experts" formulating them. These explanations simply didn't hold up in the field, which was no surprise, since they were so arbitrary. Apparently, the authors of such theories described things the way they wanted them to be, instead of admitting their ignorance. Geofantasy is well illustrated by the standard explanation of how mountains are formed. Western geologists hold that lateral pressures force mountains up, but Soviet geologists maintain the contrary—that vertical pressures force mountains up. Lateral pressure or vertical pressure—which is it? Public confidence in science's understanding of mountain building is not fortified by such contrary views.

Another case of geofantasy involves continental-drift theory, which in essence states that the earth's crust is made up of a dozen huge plates that move over the planet's molten interior. Before 1965 the theory of continental drift was considered so much nonsense, but today nearly every geologist endorses it. This is quite a sudden change of opinion, especially when we now believe that the collision of drifting plates contributes to the production of earthquakes. Yet the theory of continental drift was formulated in 1912 by the German meteorologist Dr. Alfred Wegener, who was not afraid to recognize the many available clues for what they were. But no one took him seriously then. Why? It seems that he was psychically oriented. After he was seen dowsing a fault from the back of a yak at an international geological conference field trip to the Urals in Russia, prejudice in the scientific community acted swiftly to undermine Wegener's standing. He lost his credibility and his theory was forsaken.

When I first read Cayce's and Nostradamus' earth-change predictions I was fascinated. But I paid them no more mind than I would any Jules Verne-like peek into the future. But my fascination soon deepened when I became a graduate student in archaeology at the University of Arizona because I became involved with Aron Abrahamsen, an aerospace engineer turned psychic. Abrahamsen used his psychic ability to help me make a major archaeological discovery in Flagstaff, Arizona, probably the earliest site of man in North America to date. Crude stone tools such as hide scrapers, knives, and saws were found at the exact location and at the exact depths Abrahamsen had predicted during the special psychic state he enters. Some were up to 30 feet in depth and over 100,000 years old. This excavation and the controlled experiments we conducted together were the subjects of a paper I presented at the Seventy-third Annual Meeting of the American Anthropological Association in Mexico City in November, 1974. This work is discussed at length in my first book, *Psychic Archaeology: Time Machine to the Past*.

In the course of this archaeological work, Aron and several other psychics also demonstrated an ability to make accurate geological predictions. These other psychics included Bella Karish, who works closely with Dr. Wayne Guthrie of Los Angeles and who is noted for her medical skills, and Beverly Jaegers of St. Louis, whose success in helping police departments solve murders has been impressive. Let me give you some examples. Of thirty-four documented geological predictions concerning the type of strata the Flagstaff test shaft would encounter at each level, Abrahamsen got thirty-two correct—a remarkable accuracy record of 94 percent. At a depth of eight feet he predicted a one-foot bed of fine-grained silt would interrupt the bouldery jumble that dominated the strata. Our excavation showed it did exactly that. At fifteen feet he predicted a date of at least twenty thousand years in age for the deposit to be encountered. A radiocarbon test performed by Teledyne Isotopes fixed a date of approximately 25,470 years in age. It was predicted that at twenty feet we would encounter a very rare geologic deposit called a *paleosol*—a fossil soil—and we did. Library research showed that the other psychics' complex and detailed geologic history of the area was also accurate. This history included mountain building, the intrusion of molten rock within the mountain, volcanic activity, glacial activity, flooding, and even faulting with associated earthquakes.

If these other psychics could perceive the past so accurately, how well, I wondered, could they do with the future? How accurately could they predict future geological events such as earthquakes and volcanic eruptions? I couldn't resist testing them. Perhaps, I felt, they could comment on Cayce's predictions and fill in the many blanks Cayce left. Unfortunately, a trained geologist never questioned Cayce. But with Aron Abrahamsen, Bella Karish, and Beverly Jaegers, perhaps I could do something even more important than collecting detailed predictions. Perhaps I could get them to give information about the actual physical mechanisms and reasons behind such catastrophes. In their psychic states, several of the many seers I came to work with quickly demonstrated a sophisticated knowledge of the complex physical parameters involved in such basic geological processes as mountain building, volcanism, and continental drift—a knowledge they didn't have in their normal state of consciousness. The accuracy of this information could be tested. One needn't wait to see if their predictions came true to get a sense of their validity. If we had correct information about why such catastrophes occur, we could even make our own predictions.

Not surprisingly, in their predictions the psychics' consensus spoke of California being racked by repeated earthquakes and land falling into the sea. Like Cayce, they had the west coast of the United States eventually moving as far inland as Kansas and Nebraska. They also had the eastern states being hit by a flurry of quakes. Major climatic changes and tidal waves were also described. And all this activity was said to culminate in the year 2000 with the shifting of the earth's rotational poles. In summary, the different psychics I worked with foresaw the same general sequence of events that Cayce and Nostradamus indicated would precede this catastrophic period: first tidal waves hit India, next certain key volcanoes erupt, most of Japan being inundated, followed by California undergoing major losses of land to the sea. Ominously, on January 1, 1976, a related and highly specific prediction made by Cayce in 1932 came to pass when Mount Etna erupted and the New Hebrides Islands in the South Pacific—located directly on the other side of the planet from Mount Etna—were struck by a strong quake (6.8 on the Richter scale). *This rare simultaneous occurrence can be seen as starting the countdown for fulfillment of Cayce's literally earthshaking prophecies.* On April 9, 1932, Cayce was asked when changes in the earth's activity would begin to be apparent. Cayce answered, saying, "When there is the first breaking up of some conditions in the South Sea (that's South Pacific, to be sure), and those as apparent in the sinking or rising of that that's almost opposite same, or in the Mediterranean, and Etna area, then we may know it has begun."

To date several of Abrahamsen's specific earthquake predictions—all of which have been documented—have come to pass.* For example, in 1973 Abrahamsen predicted that in December, 1975 (± 3 months), an earthquake of approximately 4.5 on the Richter scale would strike the Los Angeles area. And indeed on January 2, 1976, an earthquake of approximately 4.2 on the Richter scale struck there. (Be it noted that his prediction of damage to the Oakland Bay Bridge in 1973 as a result of earthquake activity was a clear miss.)

More impressive is the accuracy of the (even more specific) earthquake predictions of Clarisa Bernhardt, a psychic with whom I have also worked. For example, geophysicist Dr. John Derr, coordinator of the U.S. National Earthquake Information Service in Denver, reported that on May 31, 1976, Bernhardt "predicted that there would be an earthquake in the Western Pacific on June 26. She stated that it would register 7.0 on the Richter scale." He went on:

> The prediction was logged in our computer because we are researching the accuracy of psychics in forecasting earthquakes.
>
> At 4 A.M. on June 26th, an earthquake occurred on the island of New Guinea, and it registered 7.1 on the Richter scale.
>
> Mrs. Bernhardt's accuracy was remarkable—far beyond the possibility of chance.

*These predictions were privately printed and distributed in an edition of several hundred copies. They are now out of print.

Dr. Derr is quick to add that "we work for the government. We are totally unable to endorse Clarisa in any way."

In a similar recorded case, reported in the *San Francisco Chronicle*, Bernhardt has even accurately predicted the occurrence of an earthquake to the minute. Intrigued by Clarisa's series of seven accurate public predictions, Dr. David Stewart, director of the prestigious MacCarthy Geophysics Laboratory at the University of North Carolina, put Clarisa through a series of controlled tests involving the locations of faults. She passed with flying colors. Dr. Stewart was moved to state, "I'm convinced psychics can predict earthquakes—Clarisa Bernhardt has proved that... Her accuracy was absolutely incredible!"

Just as important as specific predictions are the surprising discoveries I made in my personal odyssey as I worked with different gifted psychics and researched their statements. A light bulb seemed to go on in my head the day I realized that the drastic earth changes they had all been detailing for the United States generally coincided with the official USGS tectonic map of North America. A geological tectonic map denotes fault zones and volcanic areas. It also reflects the configuration of contours of an area's basement rocks as opposed to the configuration of an area's land surface (as reflected by a topographic map). Remarkably, the great land changes the psychics talked about, such as a new western coastline, matched up with the fault patterns and basement contours of the official USGS tectonic map.

I became more of a believer in psychic insights for the future on a day in 1973 when I learned about John Wilkins. At the beginning of one research reading with Abrahamsen, he said that a gentleman named John Wilkins appeared to him. Abrahamsen said Wilkins wore a pilgrimlike collar and had rolled-up sleeves. Wilkins told Abrahamsen that in his day he was interested in the relationship between earthquakes and the planets, and that he "came by" to help. In the reading that followed, Abrahamsen indeed spoke about such planetary relationships. I didn't know what to make of all this, and neither did Abrahamsen. Neither of us had ever heard of John Wilkins. But the next day, after a trip to the library, my former wife Phyllis returned with a pleasant surprise. From *Who's Who in Science* she learned that a John Wilkins lived in England during the middle 1600s. Interestingly, this time period seems consistent with the style of dress Aron described. Moreover, Wilkins received a doctorate in engineering from Cambridge and was a natural philosopher. He was married to Oliver Cromwell's sister, was a master of several colleges, and was a fellow and cofounder of the prestigious Royal Society. Wilkins even wrote a number of books, and here the plot thickens. These books included *The Discovery of a World on the Moon, A Discourse Tending to Prove Tis Probable Our Earth Is One of the Planets, Mathematical Magic,* and *An Essay Towards a Real Character and a Philosophical Language*.

So not only was there a John Wilkins who lived in the proper time period, but he worked in astronomy, and from the titles of his books, he would seem to have been interested in studying the planets.

John Wilkins prepared me for the shock I had when I learned that there indeed was something to Cayce's and the other psychics' statements about how the planets were involved in the triggering of earthquakes. The research of certain present-day scientists already had substantiated some of these relationships. For example, Dr. R. Tomaschek, writing in the geophysics section of *Nature*, the international journal of science published in London, said that "...a remarkable correlation between the positions of Uranus and the moment of great earthquakes has been established...."

After all of this it came as no surprise when Dr. William Kautz of Stanford Research Institute in Menlo Park, California, called in May of 1974 to say that he had learned of my work from others at SRI and wondered if he could run the earthquake data I had collected from psychics on the SRI computer. Dr. Kautz explained that the particular planetary configurational information Abrahamsen spoke of was quite similar to already-established correlations between these same planetary configurations and solar storms resulting in sunspots. It was no surprise either when Dr. Kautz excitedly called again to ask if he could work with Abrahamsen directly to get more information regarding the mathematical relationship between the planets and earthquakes.

In 1974 two scientists, Drs. John Gribbin and Stephen Plagemann, predicted in *The Jupiter Effect* that the coming rare alignment of planets in 1982 "will trigger off regions of earthquake activity in Earth." Gribbin is an editor of *Nature* and holds a doctorate in astrophysics from Cambridge University. Plagemann, who holds a doctorate in physics (also from Cambridge), is a researcher with NASA. While at Cambridge, he worked at the Institute of Theoretical Astronomy under the astronomer Sir Fred Hoyle.

Psychics are not the only ones able to predict earthquakes. The importance of the unusual behavior of animals as an indicator of impending earthquakes cannot be overemphasized. In 1976 the Chinese government's Earthquake Office issued an illustrated pamphlet detailing various types of erratic behavior in animals that can be used to tell when an earthquake is imminent. In fact, both Chinese and Japanese seismologists have recently used zoo, farm, and laboratory animals to predict earthquakes. When snakes came above ground and froze in the winter weather, and when farmers reported strange animal behavior in the vicinity of Haicheng, Chinese officials had the city evacuated. A few days later, on February 4, 1975, a strong earthquake (7.3 on the Richter scale) destroyed about 50 percent of the city. Today American biologists, animal behaviorists, and geologists from the USGS and such prestigious institutes as Scripps Institute of Oceanography and Stanford University are actively studying the ability of animals to warn us of earthquakes.

Soon after I finished my initial studies of psychically derived earthquake forecasts in 1973, I was amazed to read an article in *Science* by Columbia University geophysicists concerning rock dilatancy, a promising new physical basis for earthquake prediction. The Columbia geophysicists stated that they were able to successfully predict a number of earthquakes on the basis of the tiny cavities that form in rocks just prior to their fracturing, a subtle change also involving water saturation that can be picked up by seismologists.

The rock dilatancy and changes in water saturation that the Columbia scientists wrote about essentially detailed the same physical changes preceding an earthquake that Abrahamsen had just related to me in a more personal idiom. Instead of rock dilatancy, Abrahamsen spoke of the formation of "voids." Instead of changes in water saturation, Abrahamsen spoke about changes in the water table. Today, the dilatancy principle heads the list of most scientific earthquake predictive factors, which also include crustal "swelling," crustal tilt, changes in the local magnetic field, changes in electrical resistance of rock, changes in the radon (a radioactive gas) in a quake area, and even changes in the local water table, as Abrahamsen noted. In fact, in 1975 Chinese seismologists used water-table changes as reflected in wells to successfully predict several quakes. Unfortunately, in the United States today only a small number of stations have been set up to study and monitor these physical earthquake indicators. We will look more closely at this topic in Chapter 5.

While Abrahamsen psychically reviewed the various physical factors and mechanisms involved in earthquakes, he named one factor that scientists had totally missed—temperature! Abrahamsen told how the temperature of the strata deep below the surface begins to rise just before a quake, first gradually and then very rapidly, and the bigger the temperature rise, the bigger the quake. Abrahamsen even gave an exact formula for measuring temperatures at different depths and computing the severity of the quake and the number of days before it will strike.

According to the laws of physics, Abrahamsen's temperature predictor makes good sense and it is amazing that scientists haven't come upon it. In terms of economics, temperature would probably be the simplest and cheapest quake-related factor to monitor. It also potentially allows scientists to compute the exact day a quake will strike. This could be invaluable in terms of orderly preparation and evacuation before an impending quake.

Bella Karish, the most science-oriented of the psychics I worked with, believes with many geologists and geophysicists that the earth has two cores, an inner and an outer. Like those same geologists and geophysicists, she believes that activity within the cores is a critical determinant of most geological events on the earth's surface—from earthquakes to volcanic eruptions, to openings of ocean rifts, to continental drift, to mountain building. But unlike most geologists and geophysicists, who are unclear as to the exact mechanisms involved, Karish in her trancelike psychic state readily offers a number of provocative theories, using technical language that one wouldn't expect from a person who never went to college and whose reading, she claims, is confined to spiritual literature.

In addition to temperature and internal earth mechanics, the psychics also indicated a number of other factors that could be used for earthquake prediction—factors which, like temperature, scientists have overlooked so far. I was dazzled by the way they seemed not only to see into the future and make predictions about earthquakes and earth changes, but also by the way they were apparently able to view the physical processes working behind such events.

Probably the most extraordinary of the many insights I obtained was this: that human social activity and even human thoughts played a role in the occurrence of earthquakes and climatic changes! Cayce often talked about the production and activity of "thought forms." He related what I here call a biorelativistic theory (Chapter 9) that brings to mind a number of biblical themes and even the rain dance of some American Indian tribes. Indeed, all the psychics spoke about how the drastic earth changes predicted could be mitigated via thought forms of biorelativity.

At first I didn't know what to make of such information, so as part of my graduate studies I conducted some controlled experiments to test this "mind influencing matter" concept. I didn't know who was more shocked—my professors or me—when the results supported Cayce's contention. The Cayce readings constitute an encyclopedia of information and clues that can take us many steps closer to a rational understanding of and ability to predict earthquakes. Now biorelativistic factors could be added to the physical, geological, and planetary factors he pioneered in identifying. And what's more, changes in all four areas are consistent with the great number of earth changes he and other psychics have predicted for the near future, as I'll show in later chapters.

For the first time I was overwhelmed with the feeling that the predictions of Cayce and Nostradamus, reemphasized and detailed by the various psychics I worked with, might really come true and that we might realistically expect to undergo a generation of global catastrophes. Even if everyone ignored the psychics' many predictive indicators, I felt lives could be saved if the particular sequence of events that they all agreed was to precede and signal this destructive decade was actually to occur. That sequence should be enough warning for the most resistant skeptic that a period of greater cataclysm was imminent. And the psychics' mind-boggling predictions for the years 2000–3000 A.D. would take on a new significance and meaning. These predictions sound very much like the millennium promised in the Book of Revelation. They all movingly speak of the coming (return?) of a messiah. Some even predict a series of events that can be followed up to this event. We will look at those predictions in Chapter 11. But before we go ahead, let's see what there is to learn closer at hand, remembering that psychics are only the most vocal of a number of different groups warning of a generation of earthquakes and cataclysms. Surprisingly, a number of geologists, geophysicists, paleoclimatologists, archaeologists, astronomers, meteorologists, biologists, animal behaviorists, and students of the Bible speak of the probability of an upcoming destructive period. With the stakes measured in human lives so high, we can't afford to ignore the possibility of an earthquake generation, especially since there seem to be ways to avert, mitigate, or avoid the cataclysmic changes indicated. At the very least we must increase our awareness of the very real danger of major quakes whenever or wherever they come.

CHAPTER 2

The Predictions: 1980-2000 A.D.

The elderly woman sitting across from me on the Greyhound bus kept glancing at the long cardboard tube I had leaning against the window. From the anxious expression on her face, it seemed that she thought I had a rifle or shotgun in the tube and that I was soon going to hijack the bus. I decided to put her fears to rest, so I explained that I had a number of large geological maps rolled up in the tube—maps that were needed for research when the bus deposited me in Oregon. She nodded and said she was from San Diego, and was going to visit her children in San Francisco. She felt relieved. But little did she know I planned to research something that could give her cause to worry like she'd never worried before!

I was on my way to visit Aron Abrahamsen and conduct a series of readings on future earth changes and the basis for these changes. Abrahamsen lived in Medford, Oregon, at the time. As the bus sped along, I couldn't keep my mind off Edgar Cayce's cataclysmic predictions for the future. I was excited about being able to examine Cayce's readings through a competent psychic, thereby attempting to amplify them.

In his lifetime Cayce, a professional clairvoyant with only a seventh-grade education, made more than fourteen thousand stenographically transcribed readings. These readings are preserved and made available to the public by the Association for Research and Enlightenment (ARE) in Virginia Beach, Virginia, the organization built around Cayce's work. Fifteen of these readings concern events predicted for the period 1958–2000 A.D. and they provide a base line against which the predictions of present-day psychics should be measured. The readings relate a pattern of increasingly widespread natural and social catastrophes. For example, in 1934, Cayce in his sleeplike psychic state said:

> As to the changes physical again: The earth will be broken up in the western portion of America.
>
> The greater portion of Japan must go into the sea.
>
> The upper portion of Europe will be changed in the twinkling of an eye.
>
> Land will appear off the east coast of America.
>
> There will be upheavals in the Arctic and in the Antarctic that will make for the eruptions of volcanoes in the torrid areas and there will be the shifting then of the poles—so that where there have been those of a frigid or semi-tropical will become the more tropical, and the moss and fern will grow.
>
> And these will begin in those periods in '58 to '98 when these will be proclaimed as the periods when His light will be seen again in the clouds.

The reading was given on January 19, 1934, and is identified in the ARE index code as 3976-15. However, this reading, like the others, was frustratingly general. *How* would Arctic upheavals "make for the eruption of volcanoes in the torrid areas"? I wanted more specific locations and dates. In only one reading did Cayce predict a specific event at a specific time. Appropriately, this concerned the "western portion of America" being "broken up." In 1936 Cayce said, "If there are the greater activities in the Vesuvius or Pelée, then the southern coast of California—and the areas between Salt Lake and the southern portions of Nevada—may expect, within the three months following same, an inundation by the earthquakes" (270-35, January 21, 1936). But no data were given concerning these initiating eruptions. From a reading Abrahamsen had done for a friend of mine, I already knew that he was in essential agreement with Cayce and was prepared to give specific information about the earthquakes and the land falling into the sea, which he also foresaw for California. I was understandably eager to get on with it.

The bus wound its way northward, quietly slipping in and out of California valleys and basins that in the remote past had been filled by the ocean. The Pacific had once swept over the area and held sway. I could almost feel how it would return and lay claim to its former holdings. I wondered what would become of the people living here if the ocean did in fact wash over this area again.

It was December, 1972, and I had a one-month vacation between semesters at the University of Arizona, where I was working on my master's degree in anthropology. I planned to spend the entire month working with Abrahamsen on earth-change readings. Who was this extraordinary man?

Aron Abrahamsen, now in his late fifties, came to this country from Norway at the age of eighteen. He served in the U.S. Navy during World War II and afterward earned a degree in electrical engineering from California State Polytechnic University. He went to work in the aerospace industry. At one of the leading aerospace companies, he was a member of the first special ten-man scientific team appointed to study the feasibility of going to the moon, long before the Apollo project came into being.

Like Cayce, Abrahamsen is a deeply religious man. He is also something of a biblical scholar, equally at home in Hebrew and Greek. In 1962 he began a weekend retreat center at Big Bear Lake in southern California. In 1970, after having phased himself out of his aerospace career, he became an ordained minister in the Universal Christian Church.

About that time Abrahamsen found that in his daily meditations he could "tune in" on the past, present, and future lives of people he was counseling. He found, to use his own words, that he had access to information from a "higher source" than himself. He began to do so in order to get a better understanding of his clients' situations. Having read about the work of Edgar Cayce, Abrahamsen wasn't too surprised by the psychic ability that rapidly flowered in him. A ministerial associate soon put his ability to the test, and with his psychic insight Abrahamsen was able to help in several critical medical cases. Soon his day was filled with people who came to him for psychic readings on a wide range of subjects.

By 1974 Abrahamsen had done more than 1,500 readings for individuals, 985 of whom he never met. These personal readings offered individual counseling—both psychological and medical—for life and work. The general response from those who have received the readings is one of gratitude for the helpful, fatherly understanding given them. I personally know a number of people who have had readings from Abrahamsen, and I have seen his insight and advice work out for them.

Abrahamsen's psychic perceptions are not equivocal. He *can* be pinned down. For example, a group of doctors at the ARE Medical Clinic in Phoenix, Arizona, a research clinic with vast experience in epilepsy, recently had a young epileptic whom they could not help at all. The girl was having fifty to sixty grand mal seizures a day. In desperation the doctors advised the mother to seek the aid of a psychic. Abrahamsen's name came up. The mother contacted him, and he gave an emergency reading for the girl that suggested specific unorthodox treatments. These treatments were followed and in just two weeks the doctors considered the girl almost totally cured. Her condition had improved to the point where she was having only one or two mild seizures at night in her sleep. The doctors were so impressed that they advised further readings to explore the reasoning behind the unusual treatments. They are also beginning to refer other patients to Abrahamsen.

There are some important differences between the way Abrahamsen and "sleeping prophets" such as Cayce operate. Unlike them, Abrahamsen recalls what he has said and can respond to follow-up questions that seek further details. He is also capable of giving psychic readings that analyze his own previous statements—which is just what I had him do.

I learned of Abrahamsen in 1971 from friends at the ARE. Hugh Lynn Cayce, Edgar Cayce's son, had come to test Abrahamsen and thought highly of him, as did the many others at the ARE who had received readings from him. They felt he had the same psychic capabilities as Edgar Cayce, a comparison they are careful in making. At this time I began to test Abrahamsen's ability by employing his help in my archaeological research. This resulted in the discovery noted in Chapter 1.

Soon after Abrahamsen picked me up at the bus depot we went over the outline and the tentative schedule. We went at it just like two engineers would on any project. We had to work in the earth-change readings among the counseling readings that Abrahamsen already had scheduled.

I didn't want to alarm anyone with the information that might come out. My purpose was simply to learn, from the vantage point of the psychic states of Abrahamsen and others, what might happen and then to ascertain whether there was any scientific basis for such possibilities. There was no way to assess the accuracy of what was said except by the passage of time. Exact locations and dates given could only be viewed as guesses, and we agreed that the main value of such exact information was for the purpose of simplifying communications between the psychics and myself. Only the future could assess Abrahamsen's accuracy. By making the information available to people long before the predicted events, the results would speak for themselves. But we were keenly aware that what is a sufficient warning for one person may not even draw the attention of another.

The next day Abrahamsen, his wife Doris, and I joined our hands in prayer, asking especially that our undertaking would be of help to our fellow man. Abrahamsen sat quietly in his favorite chair. In a few minutes he had entered his meditative state.

Then I asked him to speak on the initial subject: the first earth changes expected. I interrupted him when clarification or amplification was needed. At my side were my notes and geological maps. It didn't take long for me to have my maps spread all over the floor of the room we were in, and as Abrahamsen spoke, I scurried about, referring to them. There were tectonic maps, depositional maps, topographic maps, and road maps for the United States and for the world.

I hardly had to ask a question aloud; Abrahamsen seemed to pick up my questions and answer them as soon as I had formulated them in my mind. Moreover, as I marked the spots on the maps, even though his eyes were closed and he faced forward as usual, and even though I was often at the far end of the room, Abrahamsen showed that he knew where my pencil was because he would tell me to move it an inch or two this way or that to a particular town or city that he would then name.

This clairvoyant map reading was remarkable in itself. But it wasn't nearly as dramatic as the key predictions for 1980–2000, including year-by-year changes in configuration of the western coastline of the United States. Later, the readings were transcribed and Abrahamsen published them privately under the title *Earth Changes*. Even though I thought a great deal of his ability and was impressed with his detailed readings, I still had to be careful, remembering that Abrahamsen was familiar with Edgar Cayce's readings. I had to separate what was uniquely Abrahamsen's from what constituted mere repetition or variations on Cayce's themes. I was especially concerned about this because I had heard that some researchers at Stanford Research Institute didn't think much of Abrahamsen, calling him just a "rehash" of Cayce. (Ironically, these same researchers later came to use Abrahamsen on some of their subsurface exploration projects.) Undeniably there was a great deal of overlap between Abrahamsen and Cayce, but from the unique details Abrahamsen added, it seemed he was drawing his information independently, from his own peculiar source.

As it turned out, my bus ride into the future didn't stop with Abrahamsen—it had only just begun. Soon after I became comfortable with Abrahamsen's earth-change readings, I underwent another big shock. Ironically, it was the shock of recognition of the familiar.

I had felt pleased about my work with Abrahamsen, taking a month of close contact to get detailed information on Cayce's readings. After first thinking that the details unique to Abrahamsen were extremely valuable, I then began to feel that Abrahamsen was merely a rehash of Cayce. I resolved my anxieties, however, and everything seemed to be going smoothly—that is, until I began to work with Bella Karish, Beverly Jaegers, Ray Elkins, Susan Harris, and Clarisa Bernhardt, asking them the same questions I had asked Abrahamsen. It's then that I experienced shock as I realized that all the other psychics were predicting the same earth changes as Abrahamsen.

Was there some sort of conspiracy going on, I wondered? Were the other psychics slipping into my room at night to read my notes? Were we all involved in some sort of mass hallucination? If these other psychics hadn't had their own equally impressive "track records," I wouldn't have known what to think. To be sure, there were differences among the predictions, but they were minor. The biggest difference came when one psychic had a country subsiding beneath the oceans while another had it rising, but—and this still amazes me—they all singled out the same set of trouble spots and types of change.

It is easy enough to put aside the dire predictions of Abrahamsen, Cayce, and Nostradamus as curiosities, but when a group of five other gifted psychics each independently relates the same essential story in his own unique way, then there is more than pause for thought. I wish I could explain exactly why the correspondence was so good, but to answer this question I would have to know the mechanism behind psychic phenomena—which no one knows at present. However, it was clear to me that I didn't need to know the "why" behind the correspondences to take advantage of them. When I began my work, I simply wanted to obtain testable information and to verify it.

One of the other psychics I came to question on future earth changes was Beverly Jaegers, a psychic who had previously helped me in my archaeological excavation. Jaegers' specialty is psychometry. She can touch an object and gain impressions of its history. When I first met her I gave her a small piece of nondescript bone and she quickly described a scene featuring camels and mammoths. The bone had come from a 10,000-year-old mammoth "kill" site in New Mexico. She has used her psychic ability to help police detectives solve a number of murder cases. In the dramatic kidnap-murder case of socialite Sally Lucas of St. Louis, Jaegers astounded Lieutenant John Kiriakos of the Missouri State Highway Patrol and his associates by pinpointing the location of the victim, who was buried in a dry creek bed twenty-five miles away. Each month police departments around the country send Jaegers dozens of objects from homicide cases for her and her coworkers to try their hands on. They don't solve every case, but judging from the way departments keep sending her items, she generally provides useful information.

Her psychic talents include precognition as well. Over the course of several months, a businessman periodically visited her and gave her sealed envelopes to hold, the contents of which she never knew. In one instance Jaegers described a strike scene and turmoil. After holding another envelope she said she saw a tank with a red star on it, and Chinese and Cuban soldiers fighting together. One day the businessman returned delirious with joy. He told her that papers relating to coffee futures were in the sealed envelopes and that he had been "playing" coffee on the commodity market based on the information Jaegers gave him after handling each envelope. He told her that he made a fortune based on her looks into the future. He credited her, in her earlier psychic readings, with correctly anticipating the soaring coffee prices of 1976 and the resultant consumer boycott. Later she anticipated the war in Angola, in which Russian, Chinese, and Cuban advisors guided the various forces in that strife-torn country. Few people realize that Angola is a major coffee exporter and the turmoil there greatly added to coffee shortages. To express his appreciation for her help, the businessman bought Jaegers a brand-new $60,000 house in St. Louis.

In my work with Jaegers I not only questioned her about the future, but also gave her sealed envelopes containing small amounts of soil. One envelope contained sand from the beach at Los Angeles. Another had soil from the San Andreas Fault outside Palmdale. Still another came from the volcanic area of Yellowstone Park in Wyoming. Several envelopes used as controls contained soil from Ohio. Of course Jaegers was not told what the envelopes contained. After handling the envelopes, Jaegers, in her readings, was able to tell, accurately, what part of the country the samples came from and the characterizing (dominant) geological activity for the area. This gave me increased confidence in her predictions.

Bella Karish was another to whom I came in order to learn more about the future. She too had helped me in my archaeological work and I was greatly impressed with her ability. In Chapter 1, I mentioned that Karish included in her predictions highly technical information about the interior of the earth.

I also came to question Clarisa Bernhardt. In the same chapter I noted Clarisa's amazing public record of seven consecutive accurate earthquake predictions. Months in advance she had accurately predicted the magnitude, location, and time of several major earthquakes. When I questioned Bernhardt about long-term predictions, she talked in great detail about California.

Reverend Ray Elkins had privately published many psychic predictions dealing with earth changes before I came to question him, and so my questions to him were minimal. Ray's primary talent is psychic healing; he could be called a faith healer. Although I was rather skeptical about faith healers, I have a young daughter, Joy, who was crippled with rheumatoid arthritis at the time I encountered Ray. After X-rays, fluorograms, and laparoscopy, the doctors to whom my ex-wife and I took Joy said that there was little they could do, that Joy would have to learn to live with it. On some days Joy could hardly get out of bed because her knees were almost locked. So despite our doubts, we asked Elkins to treat Joy. Three months later, after a number of sessions with Elkins, my daughter was ranked at the top of the local junior tennis competition.

Reconditioning future tennis stars has nothing to do with making accurate earthquake predictions, of course, but I soon found myself buying and reading all of Elkins' publications. Once again I was surprised to find almost the same story. That Elkins gave the same schedule of events for California, even down to drawing the same new coastal maps, was eerie, especially since I had only shown a few people the maps Abrahamsen had me draw. Elkins could not have seen one. When I finally met Elkins in 1977 and spent a day talking and working with him, I was impressed by how readily this tall, soft-spoken man who made every word count addressed himself to each question I asked—no matter how specific or technical.

Probably the most extraordinary confirmation of Cayce's predictions came from the public at large. One day in Phoenix I met Susan Harris, a psychologist whose main interest is hypnosis. She earns her living by hypnotizing people and regressing them so they can confront problems such as excessive smoking or drinking or an old emotional scar. Sometimes, however, Harris asks people to "project," to go forward in time. When Harris lectures on hypnosis she usually demonstrates by picking people at random from the audience. Many of the people picked are not sympathetic to the psychic realm or the possibility of drastic future earth changes. Yet Harris says that when she asks people at such demonstrations to project into the future and report what they see, they frequently describe California and the rest of the world as having undergone great changes. They tell of new coastlines and new lands in terms that bring the visions of Abrahamsen and Cayce clearly to mind.

This again raises the question of the source of information that these various psychics and people under hypnosis may be tapping. According to his own readings, Cayce's sub-conscious mind contacted the "Universal Mind," which suggests what Dr. Carl Jung, the noted Swiss psychologist, called the collective unconscious. Cayce also said that everyone does what he does in contacting the Universal Mind in their dreams each night. Perhaps there are a number of paths to this Universal Mind: dreaming, psychometry, hypnosis, meditation, trance, etc. And perhaps here lies part of the reason for the repetition of the theme of a coming earthquake generation. If there is such a source I suspect that as we get closer to the events, we shall all hear of increasing numbers of people who produce psychic warnings about the earthquake generation through their dreams and other altered states of consciousness.

The psychic view to be presented in this book is a distillation of predictions given by the psychics I have introduced. It will be compared to the predictions of Edgar Cayce and Nostradamus. These by no means represent the views of all who have spoken or written on this subject. I present these particular prophecies because I have confidence in them for the reasons already indicated. And most important, as we shall see in the next chapter, it is possible to subject these predictions to logical and scientific analysis. Note, however, that the Jehovah's Witnesses, the Mormons, the Stelle Community of Cabery, Illinois, and many other religiously oriented groups also have very definite ideas about what calamities the future holds—ideas not wholly congruent with the psychic view given here. And note also that many other psychic predictions exist, ranging from the writings of the German philosopher and clairvoyant Rudolf Steiner to the Virginia Beach psychic and prophet Reverend Paul Solomon. Nevertheless, by following the view I present here, the reader will have confronted the main elements of these other predictions for an approaching cataclysmic future.

Like Cayce, the psychics I worked with predicted relatively minor earth disturbances for the 1980–90 period But for the decade of 1990–2000, again like Cayce, they predicted worldwide catastrophic disturbances. The 1980–90 period seems to be a time when the earth's internal "burners" will be turned up from their normal simmer to medium heat, in preparation for the 1990–2000 period when they will be turned up to "high."

From 1980 to 1990 my psychic group foresaw major earthquakes on both the eastern and western coasts of the United States. According to Abrahamsen, there will also be inundations by the sea on both coasts. In this initial period, and so far as the West Coast is concerned, Cayce spoke only of California being "inundated" by earthquakes. The psychics predicted more specifically that the

California cities of San Diego, San Clemente, Los Angeles, and San Francisco will experience widespread destruction from quakes. As a result they foresaw the ocean making inroads on the land toward the end of this period.

I asked Abrahamsen for details of the new coastline during this initial period of destruction. He did so, and this particular point illustrates why, as I pulled together the predictions of the other psychics, I felt such shock. This new coastline as predicted by the other psychics was virtually the same. They phrased it in different ways, but the details were similar. Abrahamsen gave his information to me in December, 1972. Unknown to Abrahamsen or me, Ray Elkins began making such warnings in April, 1970. The coastline Abrahamsen gave was exactly the same as the coastline Elkins gave. Abrahamsen said the coast would cut in from Baja California and the Gulf of California and move inland. Then moving northward it would pass the cities of Pomona (east of Los Angeles), Bakersfield, Fresno, Turlock, Sacramento, and then

New coastline resulting from the first period of destruction (1980–1990) projected by the psychics.

return to its present coastal position at Eureka.

Elkins' coastline also began in the Gulf of California area, moved inland, and returned to its present position at Eureka. Elkins noted that the Imperial Valley in southern California would first slowly fill with water and that the Colorado River would change its course before land would start to split away from the coast. But instead of naming a number of cities along the new coast as Abrahamsen did, Elkins said the new temporary coast could be visualized by drawing a straight line from Eureka through San Bernardino and continuing south toward the Gulf of California. When I did this, I got quite a surprise. The line passed by the very same cities Abrahamsen had named with the exception of San Bernardino, which was just a scant twenty miles east of Pomona, the particular city in the area Abrahamsen had named. Bakersfield, Fresno, Turlock, Sacramento, and Eureka—all the other cities Abrahamsen had named—were likewise to be found along Elkins' line.

There is some disagreement in what both men foresee for the land to be lost to the sea, however. Abrahamsen sees it all going under whereas Elkins talks of the land "splitting away" from the present coastline to form the "Islands of California." In the next chapter we will see that there is a geological basis for this difference in view. In other words, sound geological evidence can be offered for both points of view. Bernhardt, the exact quake predictor, also talks of islands forming like those off the California coast today (Santa Rosa, Santa Catalina). Bernhardt says that the "state is not going to [totally] fall apart" from the seismic turmoil. Instead, the Imperial Valley will flood, putting Palm Springs under water. San Francisco Bay will be transformed into an inland sea and the cities of Los Angeles and San Diego will become offshore islands of a new continent that will rise from the Pacific floor. Elkins talks of land appearing fifteen miles off the coast of San Francisco. Bella Karish also sees islands forming off the West Coast, as do Susan Harris' hypnosis subjects. (These initial California coast predictions should serve to demonstrate the remarkable overlap within the group. For simplicity, though, from here on I will try to present the consensus as much as possible.)

The psychics foresee the Oregon and Washington coasts, unlike the California coast, remaining relatively stable during this period. However, one dramatic exception is the prediction of an arm of the sea cutting across Oregon as far east as Idaho.

Turning to the East Coast, it was predicted that New York City will be severely hit by an earthquake, which will leave gaping crevasses in its wake. New York City is supposed to split away from the mainland. The lower part of Manhattan is to sink beneath the sea. Philadelphia and Washington, D.C., are reported as receiving only minor damage. In the Midwest, the Great Lakes are supposed to increase in size as a result of land subsidence and the St. Lawrence Seaway is to be enlarged as a result of faulting.

In Canada, the southeastern cities will feel the shock waves from the large American quakes, and they will experience a large quake of their own. Canada is foreseen as undergoing minor coastal inundation, and many new inlets will be formed. The city of Vancouver might even have to be relocated. The Aleutian Islands are supposed to disappear, and land will rise in the Bering Strait. The Japan Current, which now passes by the Aleutian Islands and then down to California, is seen as shifting to a more southerly position along the California coast. Alaska is supposed to undergo major earthquakes and coastal subsidence.

The group said that the rest of the world is to receive its share of disturbances in this period (1980–90) also, but nothing as extensive as what the United States will undergo. These specific earth changes were predicted:

1. The coasts of England and Ireland will be inundated.
2. Land will rise in the Atlantic Ocean west of England.
3. The Middle East will experience severe earthquakes. Syria, Iran, Iraq, and Israel will be involved.
4. Turkey will have earthquakes and the Dardanelles Strait will be greatly widened.
5. The Black Sea will grow as a result of shoreline subsidence.

6. Greece will experience a severe earthquake.
7. Italy will also have an earthquake.
8. Late in this period the Mediterranean area will become even more active and the Mediterranean Sea will be closed as a result of land rising between Spain and Africa. (Note that Cayce spoke about the sinking or rising of land in the Mediterranean as a signal of greater changes to come, in reading 311–8, April 9, 1932.)
9. While northern Europe—countries such as Holland, Belgium, and France—will be relatively stable during this period, these countries along with Spain will feel the reverberations from the earthquakes in Italy and Greece. In particular, Ray Elkins notes that during this initial period the earth shall "stand and smoulder, for the fuse is lit from France into Israel."
10. The earth's axis of rotation will tip a few degrees.

Despite these disturbances Europe's economy will remain strong; on the other hand, the economy of the Western Hemisphere is supposed to weaken. The United States, which is to undergo the most severe disturbances during this period, will experience a general economic setback. The areas that have been devastated by earthquakes and coastal inundations will constitute a burden to the entire nation. In these areas, especially in California, people will be in shock and panic for some time. The federal government's resources will be stretched to its limits because those resources will have to be divided among so many stricken areas. The government will be able to do little beyond providing the basics of food, clothing, and shelter. Government loans to aid reconstruction will be limited and commercial credit will be especially tight. Abrahamsen says that mortgages will be foreclosed, but as in the 1930s, there will be no one to "pick up" these properties cheaply.

A severe food shortage was foreseen for the stricken areas, which will force the people affected to start growing their own food. It was advised that food and tools be stored, along with water, clothing, and seeds. Barter will be common in the stricken area, not only on a person-to-person basis but also on a community-to-

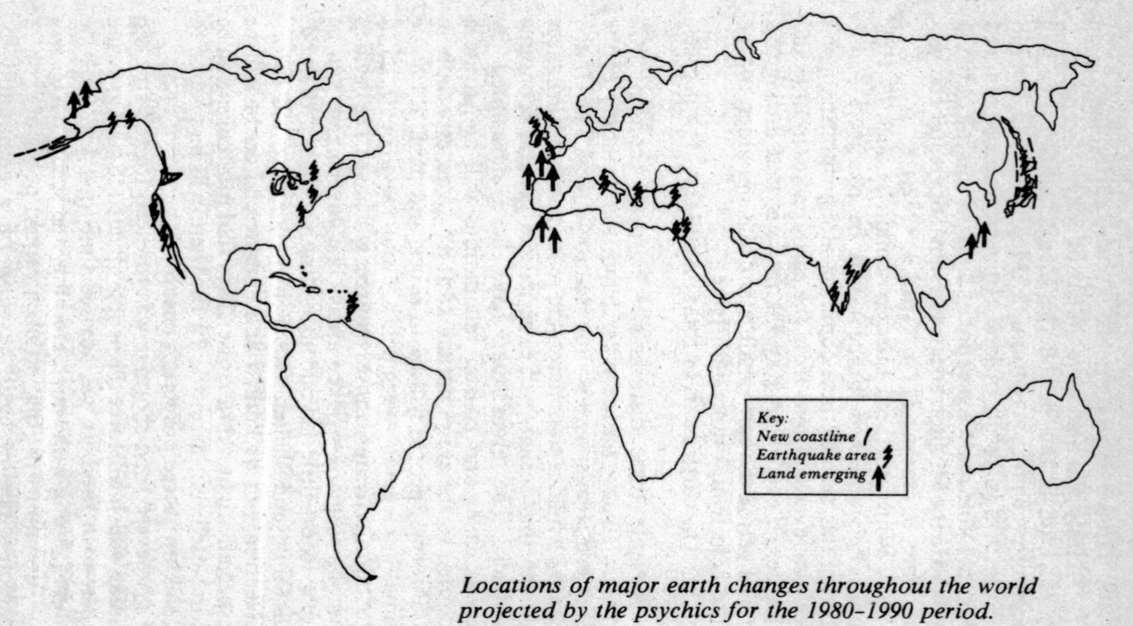

Locations of major earth changes throughout the world projected by the psychics for the 1980–1990 period.

community basis. The stricken areas will take a long time in recovering—at least five years—and many people will move to bordering states, such as Nevada and Arizona. Unaffected areas are seen as undergoing recovery pains along with the devastated areas.

The earth disturbances foreseen for this first period seem drastic enough, but they are minor compared to those foreseen for the second period (1990–2000), which I call the Decade of Cataclysm. The beginning of this more destructive decade is to be initiated by a definite sequence of events:

1. Major tidal waves and earthquakes will first hit India.
2. After a short time, the energy behind the Indian events is to "follow its path through the earth's crust," causing major tidal waves and earthquakes to ravage Japan. Major portions of Japan will go under the sea, and as a result Japan as a nation will become a minor power.
3. About this time Mount Vesuvius in Italy will have a major eruption brought on by a violent earthquake—catastrophes that will damage not only Italy but also France and even Scandinavia.
4. This activity in Vesuvius will trigger a volcanic eruption in far-off Martinique's Mount Pelée.
5. Several years after the start of this sequence (in India), the United States will be brutally hit by a new set of earthquakes, much worse than the earlier (1980–90) disturbances. The catastrophes striking the United States will become more frequent and more severe. Finally, after New York City has been completely broken up, Europe will begin to undergo major changes.

Cayce has some of these elements in his predictions. For example, he said, "The greater portion of Japan must go into the sea" (3976–15, January 19, 1934) and "If there are the greater activities in the Vesuvius *or* Pelée, then the southern coast of California and the areas between Salt Lake and the southern portion of Nevada may expect, within the three months following same, an inundation by the earthquakes" (270–35, January 21, 1936, emphasis added). In another reading, Cayce also includes volcanic activity in Sicily's Mount Etna as another indicator (311-8, April 9, 1932).

Remarkably, Nostradamus more than four hundred years ago also seems to have the same elements in his mystical predictions. Consider this quatrain:

> The Great Round Mountain of seven stadia
> [Vesuvius?]
> On the wake of peace, war, famine, and inundation
> [India?]
> It will roll a long way, sinking many countries
> [Japan and portions of California?]
> Even ancient kingdoms and their great foundations.

As indicated by the comments within brackets, this prediction—admittedly vague and open to multiple interpretations—could be seen as indicating the inundation of India by tidal waves, the eruption of Vesuvius, the sinking of Japan, and the sinking of portions of California. The following quatrains of Nostradamus may give further clues:

> Volcanic fire from the center of the Earth
> [Vesuvius or Pelée?]
> Shall cause an earthquake of the New City
> [Los Angeles?]
> Two great rocks shall have long warred against each other
> [San Andreas Fault?]
> Then, Arethusa shall redden a new river
> [Vesuvius and/or Pelée?]
> (Arethusa was a nymph changed by the Greek goddess Artemis into a stream that passed *beneath the sea* to come out as a fountain at a far distant place.)
> Garden of the World
> [San Joaquin Valley, the nation's largest food basket?] near the New City
> [Los Angeles?]
> In the pathway of the mountain fault;

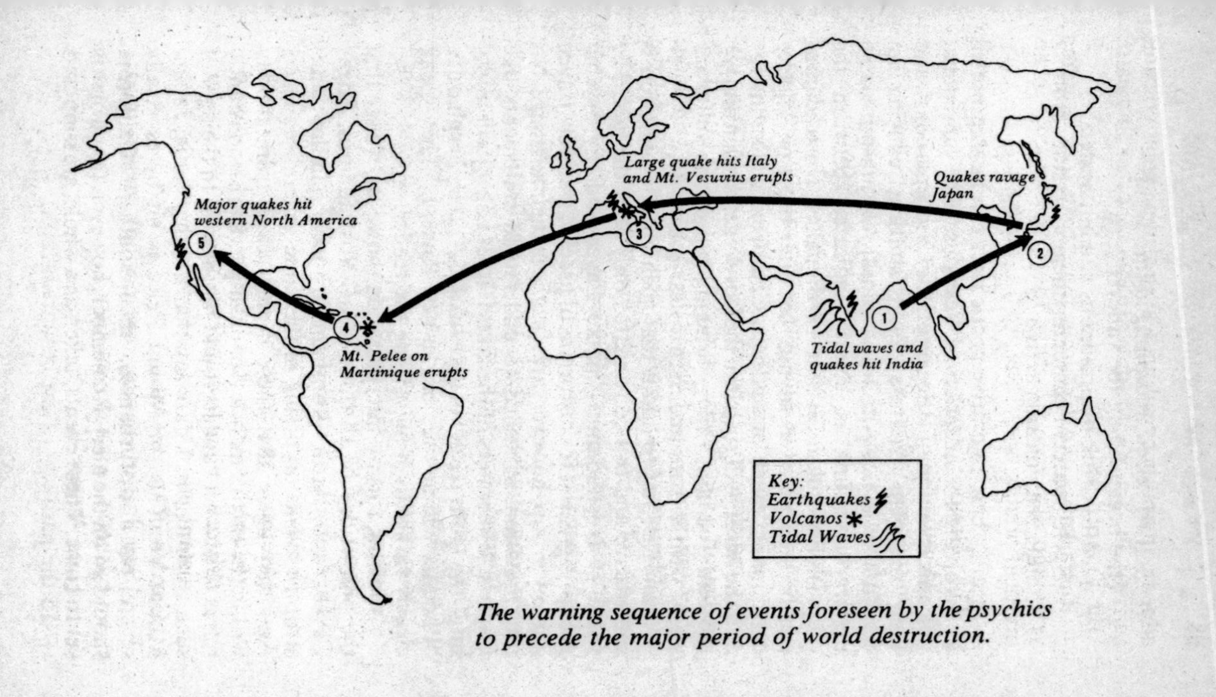

The warning sequence of events foreseen by the psychics to precede the major period of world destruction.

It shall be seized and plunged into a tub
[the ocean filling the valley?]
Forced to drink of poisonous sulphur waters.
(Such waters characteristically result from undersea volcanic activity.)

Again, these figurative poetic lines can be interpreted as involving the eruption of Vesuvius, with subsequent volcanic eruption of Pelée or off the California coast, and then the San Andreas Fault shifting, causing an earthquake in Los Angeles. A causal relationship via an underground link is clearly indicated between the eruptions of volcanoes and earthquakes. Nostradamus can be interpreted as saying that after this Los Angeles earthquake, much of southern California will submerge and the ocean will rush into the Los Angeles basin and/or the San Joaquin Valley, bringing poisonous sulfurous water with it from undersea volcanic activity.

I am the first to admit that such an interpretation is speculative and can be validated only after the fact. The value of Nostradamus, unlike Cayce and my psychic group, is greatest for historians of psychic research. Cayce and my group, however, are specific in their language. No interpretation is needed to understand statements like Cayce's "The waters of the lakes [Great Lakes] will empty into the gulf [Gulf of Mexico]..." and Bernhardt's "Phoenix will be on a bay, a new Riviera." They speak clearly rather than mystifying the reader.

Allowing for my interpretation, however, we see that the predictions of all three men—Nostradamus, Edgar Cayce, and Aron Abrahamsen—have similar catastrophic events occurring in the same general sequence. And they agree that these events result from activity within the earth's mantle (for example, Karish constantly refers to distinct "underground passages"). In the light of such a consensus, I dare say that if either India, Japan, Mount Vesuvius, or Mount Pelée is hit by especially severe earth disturbances, then southern California should go on the alert. Remember, however, that worse is yet to come. These events, taken as a whole, are supposed only to introduce the more destructive Decade of Cataclysm.

In the Decade of Cataclysm, Abrahamsen said, the overall ratio of land to water will be maintained. As one land mass is sinking, another will be rising. The changes in this decade will occur in the same general areas as the changes of the previous decade, except on a much larger scale. It is as if the changes predicted for the previous period (1980–90) constitute a precursor—a dress rehearsal, so to speak—for the changes of this more catastrophic decade.

A single earthquake of a magnitude of 8.4 releases about as much energy as was released, on the average, each year during the first half of the twentieth century. The 1906 San Francisco earthquake of magnitude 8.3 released energy equivalent to that of one million medium-strength atomic bombs. The largest instrumentally recorded earthquakes were of magnitude 8.6 (two instances), and judging from the reported effects the 1755 Lisbon earthquake, which took sixty thousand lives, had a magnitude of 8.7. The Richter earthquake scale, devised by the renowned seismologist Dr. Charles F. Richter, is a logarithmic scale based on vibrations registered on seismographs. Each whole number represents a tremor ten times more severe than the preceding whole number. That tenfold increase in tremor magnitude represents about a thirtyfold increase in vibrational energy. Thus while the tremors of the 8.5 magnitude 1964 Alaska quake were two-tenths larger than the 8.3 magnitude 1906 San Francisco quake, it was twice as severe and it released six times the energy.

A magnitude of 10.0 is usually considered the top of the Richter scale. A 10.0 earthquake would release about forty-five times more energy than the history-making quakes of 8.6 magnitude. However, the psychics indicated that for the Decade of Cataclysm, magnitudes of 11.0 or even 12.0 (almost 5,000 times stronger than the 1964 8.4 Anchorage quake) might have to be added to the Richter scale. The spirit source Elkins calls Odka said through him: "We shall say unto you, on this earth soon you shall not think in terms of 8 points upon your Richter Scale, but 10, then 12."

How could we begin to imagine the energy released by such large-scale quakes, much less the destruction that could be wrought? In 7.0–8.0 magnitude quakes, masonry, buildings, and bridges are severely damaged. Broad fissures open in the ground. Underground pipelines are put out of service. The earth slumps and land slips in soft ground. Rails are severely bent. Between 8.0 and 8.6, damage is total. Solid land has been seen to wave and ripple. Lines of sight and level are distorted, and objects are thrown upward in the air. In the 1964 Anchorage quake, which was estimated at 8.5, Montague Island in the Alaskan Gulf was tilted up 38 feet on one end, submerged 50 feet beneath the water on the other end, and moved horizontally 25 feet. Three areas of the Alaskan ocean floor were lifted more than 50 feet. In the 1960 Agadir, Morocco, quake of (approximately) 8.0 magnitude, soundings showed that the sea floor rose as much as 3,000 feet in some places. What could a quake of 12.0 magnitude bring?

From 1990 to 2000 the psychics not only foresaw earthquakes and sea inundations on both the east and west coasts of the United States, but they also foresaw earth disturbances striking virtually every state of the union. There will be very few safe areas. The bites that the Pacific Ocean took out of California and Oregon in the previous decade will steadily increase in size, affecting the entire western coastline of the United States. The coastline will move eastward in a series of wrenching surges.

Following is the final coastline: beginning from the Vancouver area of British Columbia, it goes south to Seattle, Washington, and then inland to Sheridan, Wyoming. From Sheridan it runs to North Platte, Nebraska, and then south through the western edge of Kansas to Amarillo, Texas. Turning westward, it moves to Hobbs, New Mexico, and then to El Paso, Texas, where it turns back west to Silver City, New Mexico. Then it goes on to Phoenix, Arizona; to Yuma, Arizona; and finally returns to its present position south of San

Diego, California. Such a coastline would result in the loss of parts of the states of Washington, Idaho, Montana, Wyoming, Nebraska, Kansas, Oklahoma, Texas, New Mexico, and Arizona. Worse still, virtually all of California, Oregon, Utah, and Colorado would be under the sea, and/or broken up into islands. Phoenix in particular is to become a seaport and Bernhardt pictures that "a beautiful Riviera will stretch across the southwest United States."

On the eastern seaboard the psychics said that mid-coastal states such as North Carolina and New Jersey will be inundated. Worse yet, large portions of Louisiana and Florida are seen going under the ocean. In essence the United States will be reduced from its present three thousand miles coast-to-coast distance across to less than two thousand miles. In the Midwest, portions of Wisconsin, Michigan, and Illinois, which border the Great Lakes, are foreseen going under water. This new configuration is supposed to last for two to three thousand years.

In essence my psychic group painted the same picture for the United States as Cayce did. Cayce said:

> Many portions of the east coast will be disturbed, as well as many portions of the west coast, as well as the central portion of the United States . . . Portions of the now east coast of New York or New York City itself will in the main disappear. This will be another generation, though, here; while the southern portions of Carolina, Georgia, these will disappear. This will be much sooner. The water of the lakes [Great Lakes] will empty into the gulf [Gulf of Mexico], rather than the waterway over which such discussions have been recently made [St. Lawrence Seaway] . . . Then the area where the entity [1152—the code number for the person whose reading it was] is now located [Virginia Beach, Virginia] will be among the safety lands—as will be portions of what is now Ohio, Indiana and Illinois and much of the southern portions of Canada; while the western land much of that is to be disturbed in

this land, as, of course, much in other lands [1152, August 13, 1941].

In another reading Cayce also specifically indicated that Connecticut will be affected and that Los Angeles and San Francisco will be destroyed before New York City.

A prophetic dream Cayce had on March 3, 1936, dealt with the possible new shape of the United States. As in my psychic group's readings, the state of Nebraska had a prominent role. Cayce reported:

> I had been born again in 2100 A.D. in Nebraska. The sea apparently covered all of the western part of the country, as the city where I lived was on the coast. The family name was a strange one. At an early age as a child I declared myself to be Edgar Cayce who had lived 200 years before. Scientists, men with long beards, little hair, and thick glasses, were called in to observe me. They decided to visit the places where I said I had been born, lived and worked, in Kentucky, Alabama, New York, Michigan and Virginia. Taking me with them the group of scientists visited these places in a long, cigar-shaped metal flying ship which moved at high speed. Water covered part of Alabama. Norfolk, Virginia, had become an immense seaport. New York had been destroyed either by war or an earthquake and was being rebuilt. Industries were scattered over the countryside. Most of the houses were of glass. Many records of my work as Edgar Cayce were discovered and collected. The group returned to Nebraska, taking the records with them to study... [Cayce went on to comment that] these changes in the earth will come to pass, for the time and times and half times are at an end, and there begins those periods for the readjustments... [294–185].

The psychics said that the rest of the world is also to experience major cataclysms in this decade (1990–2000). Throughout this period earth changes will be taking place

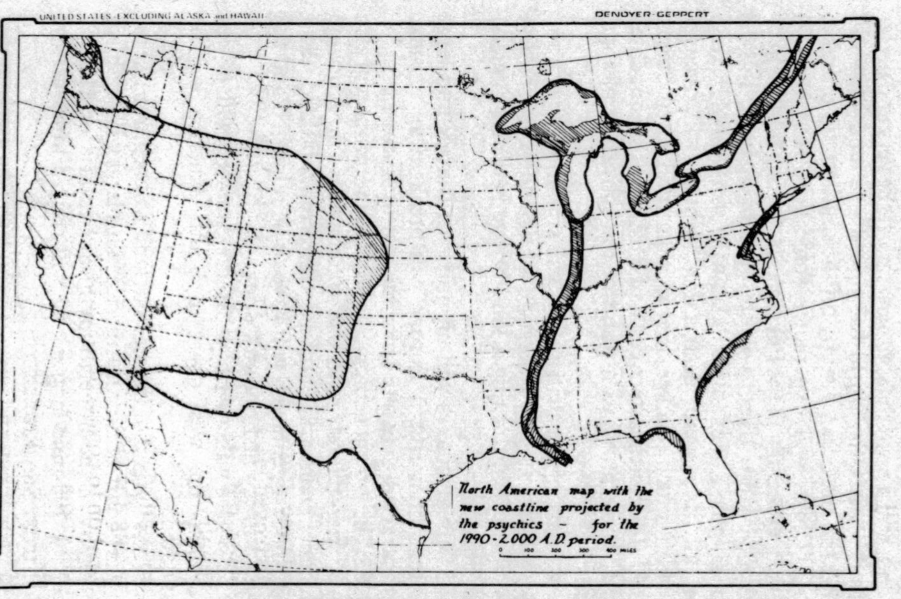

Diagonal lines designate new ocean areas.

almost constantly; we are to see "a world shaken from top to bottom." The following specific events were predicted:

1. Eastern India will lose its coast for several miles inland, and also at the southern tip. At the very end of this period—"after the world has tipped"—a land mass will appear in the ocean directly off the southern tip of India.

2. China will experience some very severe earthquakes, especially in the Peking area. Strong faults will be activated and there will be great changes in the topography. For example, lakes will form west of Peking. For all its severe destruction, China will recover quickly—much quicker, that is, than the United States.

3. The northeastern coast of Russia will be inundated (up to three miles) as a result of the activation of some faults.

4. Part of the British Isles will sink into the sea and part will rise as land emerges off England's east coast in the North Sea, forming a land bridge between England and Europe.

5. Great quakes will hit northern Europe and large portions of northern Europe will break away or sink. Denmark will separate from the rest of Europe.

6. The violent geological activities in Europe will also cause major and repeated inundations in Norway, Sweden, Denmark, Finland, England, Ireland, and Scotland. Copenhagen will have to be relocated and the ocean will lap at London's door.

7. The southern tip of South America will be cut away and the west coast of South America will be inundated.

8. At the end of this period, land will begin to rise off the west coast of South America. This land will establish the reality of the hitherto mythical Lemuria or Mu.

9. A 100-mile-long island will rise 25 miles off the coast from San Francisco.

10. Australia and New Zealand will be fairly stable during this period, despite New Zealand's active geological history.

11. Portions of Africa "shall part as though fingers have been run through it."

12. Indonesia will receive inundations. Its volcanoes, such as the famous Krakatoa, will erupt once again.

13. Most of Japan and Hawaii will be lost to the sea. However, not all psychics agreed on this, and here lie the two most notable differences among them. Karish doesn't think Japan will undergo excessively severe destruction and Bernhardt does not think that portions of Hawaii will be lost. Instead she feels that enough land will rise in the Pacific to make it possible to drive to Hawaii.

14. There will be great changes in the Arctic, Greenland, and the Antarctic.

15. Except for the cities of Vancouver, Quebec, Ottawa, and Montreal, Canada will not undergo much change.

16. And finally, this period will end with the earth flipping over, followed immediately by dramatic global changes in climate.

As was the case for the United States, the psychic group's predictions contain many of the same cataclysmic predictions for the rest of the world as Cayce's. Cayce said that "the upper portion of Europe will be changed in the twinkling of an eye" (3976–15, January 19, 1934). This seems to agree with my psychics' predictions of cataclysmic changes in the Scandinavian area and in England. Cayce even talked about "what is the coastline now of many a land will be the bed of the ocean. Even many of the battlefields of the present [1941] will be ocean, will be the seas, the bays, the lands over which the new order will carry on their trade with one another..." (1152–11, August 13, 1941). Cayce's predictions that "there will be open waters appearing in the northern portions of Greenland..." (3976–15) and that "there will be upheavals in the Arctic, Greenland and Antarctic" are in agreement with the newer predictions that there will be great changes in the Arctic, Greenland, and the Antarctic. Cayce's predictions of "eruptions of volcanoes in the torrid areas" (3976–15) are consistent with the group's prediction of the eruptions in Indonesia. Cayce's prediction that "South America shall be shaken from the uppermost portion to the end" (3976–15) parallels the

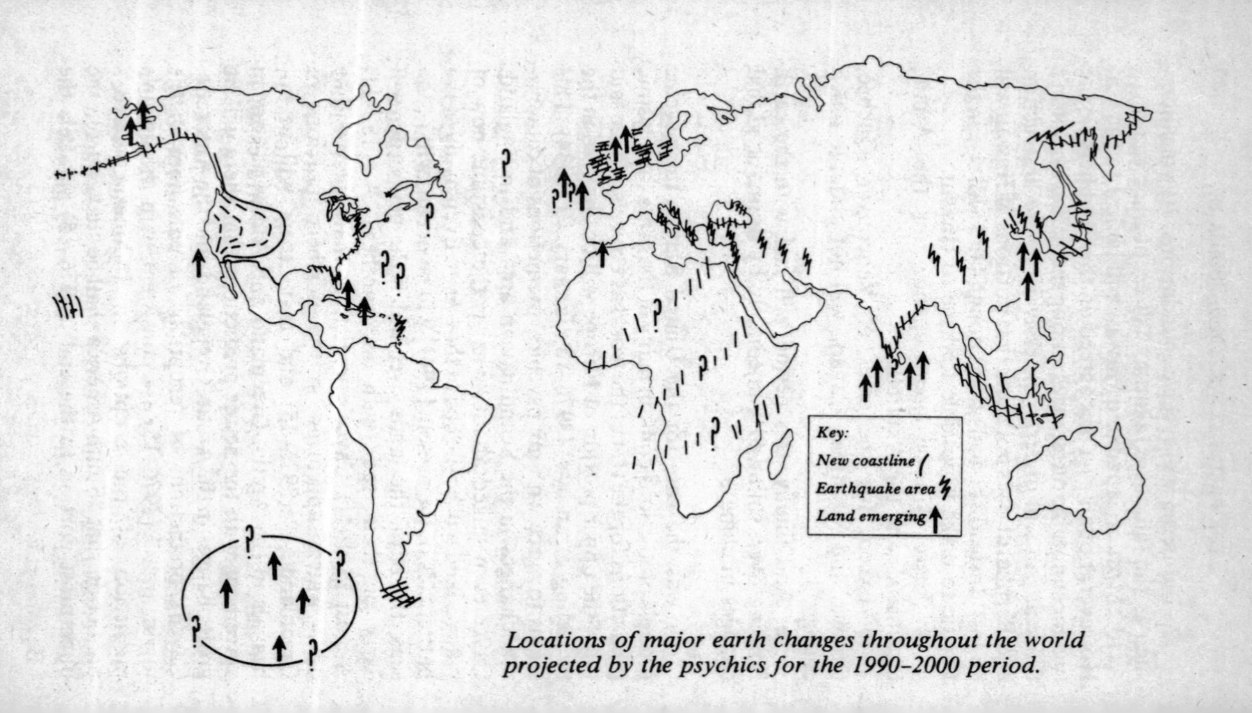

Locations of major earth changes throughout the world projected by the psychics for the 1990–2000 period.

group's prediction that the southern tip of South America will be cut away. In addition, both the group and Cayce talk of new land rising in the Atlantic Ocean and in the Pacific Ocean. And finally, all speak about the earth shifting its axis in 2000, with resulting climatic changes. In a later chapter, we'll examine the group's predictions regarding Bible prophecy, the return of "John the Forerunner," and how Jesus the Christ's light will be seen again.

During this period of cataclysm, the group foresaw, the United States will suffer severe economic setbacks. Not only will the devastated areas suffer, there are to be nationwide shortages as well. Electricity and fuel such as heating oil, gasoline, and kerosine will be in short supply. Gasoline for individual travel will be strictly rationed and used primarily for government business. Nevertheless, airplanes, trains, and buses will still operate, though on a restricted basis. As in the previous period, there will be great movements of people from stricken areas to relatively unaffected ones. Food will be at a premium and looting will be a problem. Elkins says that there will "be a time of famine when men shall fight men and snatch food from the mouths of children." There will be a great cry for jobs. Americans' idea of "necessities" will be greatly revised through this. People will have to get by on much less. There won't be "a chicken in every pot" and "a car in every garage." A long time (about thirty years) will elapse before the nation's economy recovers and flourishes again. Americans will learn the hard way that prosperity and growth are not automatic.

What about the rest of the world? Europe's economic strength up to this period is predicted to decay quickly. Europe will also have its hands full. Great economic and social changes will take place around the world, especially in South America and Africa. New Zealand is seen as the new land of hope and opportunity. Bella Karish emphasizes that on the new lands rising from the sea, and in South America, we will find "cities of gold and marble . . . with many ancient records . . . that have been hidden for many eons of time," and how these discoveries will completely change our concepts of humanity's past. Beyond this the psychics did not go.

* * *

Earth Changes at a Glance

Checklist of Major Events Predicted by Psychics

(Refer to the text for a more complete description of the events.)

1980–85

Repeated large-scale earthquakes in California, western Canada, and Alaska

The Imperial Valley in southern California starts to fill with water

Regularly occurring large-scale earthquakes, land subsidence, and coastal inundations for California, Oregon, Washington, western Canada, and Alaska

Palm Springs under water

San Diego, Los Angeles, and San Francisco destroyed

California coastline pushed back to Bakersfield, Fresno, and Sacramento

Seaway opened up through central Oregon (from west to east) reaching to the Idaho border

Vancouver, British Columbia, threatened by flood

Land rises in the Bering Strait, creating a land bridge between Siberia and Alaska

The Aleutian Islands begin to disappear as a result of volcanic eruption and earthquakes

The Great Lakes and the St. Lawrence Seaway grow larger as a result of earthquakes, faulting, subsidence, and accelerated glacial rebound

Premonitory (warning) earthquakes in the Midwest affecting cities such as St. Louis and Memphis

A major earthquake destroys a large part of New York City

Minor (warning) earthquakes in the Washington, D.C. – Philadelphia area

Land rises off the southeast coast of the United States, near Bimini in the Bahamas

Increasingly harsh weather patterns

Food shortages in the U.S.

U.S. government barely able to help all the stricken areas

1985–90

(Note: An asterisk* indicates the key warning sequence, initiating the Decade of Cataclysm.)

* Major tidal waves and earthquakes severely damage India

* Tidal waves and earthquakes severely damage Japan

* A major eruption of Vesuvius in Italy initiated by a violent earthquake that causes damage as far off as France and Scandinavia

* Mount Pelée on Martinique erupts

* The western U.S. is severely stricken by a flurry of earthquakes

New York City now completely broken up

Minor European coastal inundations

Land rises in the Atlantic Ocean west of England

Land rises near Gibraltar, creating a land bridge between Europe and Africa

Japan Current off California shifts its position

A series of earthquakes strikes the Middle East. Syria, Iran, Iraq, and Israel are most affected

Major earthquakes strike Italy, Greece, and Turkey

The Black Sea grows larger as a result of shoreline subsidence and earthquakes

European economy remains strong despite natural disasters. U.S. economy falters even further

The earth's axis of rotation tips a few degrees

* * *

1990–2000

Unusual behavior of North American animals and fish

Major disturbances (earthquakes and subsidence) on both the east and west coasts of the U.S.

Major sections of western U.S. fall into the sea as the coastline moves eastward in a series of violent surges. Final coastline established in Nebraska and Kansas

Major submergence in the Great Lakes area, with portions of Wisconsin, Michigan, and Illinois going under water

Eastern coastal states such as South Carolina and New Jersey inundated by ocean

Large quakes and coastal inundations in Connecticut and Massachusetts

Major submergence in Florida, Louisiana, and Texas

Major quakes in China. Lakes form west of Peking

USSR experiences minor coastal inundation and earthquakes

Major earthquakes in Turkey and Yugoslavia

Gulf Stream current complex undergoes a major shift

Parts of the British Isles submerge, parts rise. The ocean "laps at London's door"

Land rises in the North Sea

Norway, Sweden, Denmark, and Finland experience repeated coastal inundations

Copenhagen relocated as a safety measure

Portions of northern Europe break away or sink

Australia and New Zealand remain stable "safety lands"

Virginia Beach area and parts of Ohio, Indiana, Illinois, and the southern part of Canada remain stable "safety lands"

Land rises off San Diego

Land rises off the west coast of South America

West coast of South America inundated

Numerous earthquakes rock South America

The southern tip of South America (Tierra del Fuego) breaks away

Most of Japan gone under the ocean

Most of Hawaii gone under the ocean

Krakatoa in Indonesia erupts

Africa is torn apart by earthquake-activated faults and rifts

Great changes in the Arctic, Antarctic, and Greenland, including earthquakes, volcanic eruptions, and sudden ice buildup

Canada continues to be fairly stable, except for the cities of Vancouver, Quebec, Ottawa, and Montreal

The U.S. undergoes severe economic setbacks: food and fuel shortages, and even famine

Europe begins economic decay, leading to great economic and social change

Great economic and social change in South America and Africa

Ancient cities and ancient records found on land rising from the ocean, leading to new concepts of pre-history and human origins

John Pineal (John the Baptist) announces the imminent return of the Master Teacher

The rotational balance of the earth is increasingly compromised by polar ice buildup, changes in land-water distribution, magnetic field changes, jolts from massive earthquakes, and a rare planetary configuration

Humanity is widely warned of impending planetary earth change through dreams and visions

2000

Pole shift. Earth's axis of rotation changes locations in a sudden tumbling of the planet, during which rotation ceases for several days. The sun and other heavenly bodies appear to be fixed in their movements during this time. When the planet begins to rotate again and the day-night cycle is reestablished, the North and South poles are drastically shifted to new positions

New climates established (e.g., Alaska becomes temperate, Florida becomes cool or frigid)

New planetary and geological stability established

New Zealand becomes the new land of hope and opportunity

The return of Jesus Christ, with many helpers

Veils lifted from men's minds

Men learn their true nature and capabilities

New cultural directions

Land rises off the tip of India

Space people visit planet to observe

2030

U.S. economy regaining health

Millenniumlike conditions start to prevail

Enlightened teachers available to all who seek personal development

New emphasis by governments on helping people develop rather than on regulating their behavior and thought

Agriculture uses prayer to control rainfall and crop growth

The sun and the planet's electromagnetic field become the main source of energy

Psychics used to guide all scientific research

Medicine makes extensive use of color for healing, often with dramatic results

Cancer is cured

Limb regeneration achieved

Crystals used for healing and energy distribution/generation

CHAPTER 3

The Breaking Up of Nations: The Predictions "Hold Water"

> Before the end of this century it is virtually certain that one or more major earthquakes will occur on the North American continent.
>
> *National Academy of Sciences Task Force,* 1970

The psychics' predictions almost overwhelmed me. They weren't talking about occasional earthquakes hitting cities; they were talking about entire nations and huge sections of continents rising . . . falling . . . all around the globe, on a near-constant basis for at least a full decade. They were talking about losing the land where hundreds of millions of people presently live, and even entire mountain ranges being submerged. Through color television many people have all marveled at the spectacle and power of an oceanic volcanic eruption. But there is no way to comprehend an entire continent-sized land mass arising from the ocean's depths.

Nevertheless, mind-boggling as their predictions are, they are well within the range of geological possibility. That is not my position alone. In 1959 a professional geologist holding a doctorate in his field published an assessment of Edgar Cayce's predictions entitled "A Psychic Interpretation of Some Late Cenozoic Events Compared with Selected Scientific Data." The author, who had also written a number of research papers for scientific journals, chose to remain anonymous when his paper was published by the Association for Research and Enlightenment in Virginia Beach, Virginia, as *Earth*

Changes. The detailed analysis runs to nearly eighty pages. His conclusion? The main geological trends predicted by the psychic material "compared well" with facts accepted by earth scientists.

This evaluation would apply equally to the new psychic predictions. The main difference between the conventional geological view and the predictions of the psychic group and Cayce lies in the very rapid acceleration of certain geological processes. If Cayce and the group were right in their forecast of a Decade of Cataclysm at the end of this century, I felt somewhat reassured by my knowledge of the early-warning sequence they predicted: the Indian tidal wave, volcanic eruptions by Vesuvius and Pelée, inundation of Japan, California's loss of land. I felt that the occurrence of these events *as a sequence* was more important than knowing exact dates and places. But at the time I felt nervous because there was some scientific credibility to the predicted sequence. It's no secret, for example, that both Japan and California have lived under the constant threat of major disaster for some years. I also felt nervous because if such catastrophes were to occur, the psychics' perceptions of economic resilience were definitely pessimistic. The aggravatingly slow recovery from the 1973 Mississippi floods shows that such a gloomy post-catastrophe prognosis is not far-fetched.

After the readings on the United States were almost all in, I began evaluating them. I spread out my maps and began to contemplate them. But no matter how I pondered, the prediction that the west coast of the United States would be moved to Nebraska was too bizarre even to consider. Yes, I thought, in the United States most of the areas that the psychics said would be hit are identified as high-seismic-risk areas by geophysicists. Besides the states of California and Nevada, the following areas of the country are classified as susceptible to major damage: northwest Washington State; contiguous portions of Montana, Wyoming, Idaho, and Utah; contiguous portions of Missouri, Tennessee, Illinois, and Kentucky; the St. Lawrence River valley area.

And, yes, most of these areas also have historical records of earthquakes, even such seemingly safe states as

South Carolina, Georgia, and Massachusetts. And again, yes, Dr. Samuel Hand, head of the Vertical Network Division of the National Geodetic Survey, had said, in the December, 1972, issue of *Geotimes*, "We have determined through measurement that much of the earth's crust is moving vertically—one way or the other—at slow but significant rates." Dr. Hand pointed out that the Rocky Mountains are rising, the Houston—Galveston area is sinking (five feet in twenty years), and the area between the Great Lakes and the Rocky Mountains is rising. And last of all, tidal waves, or *tsunami*, often follow

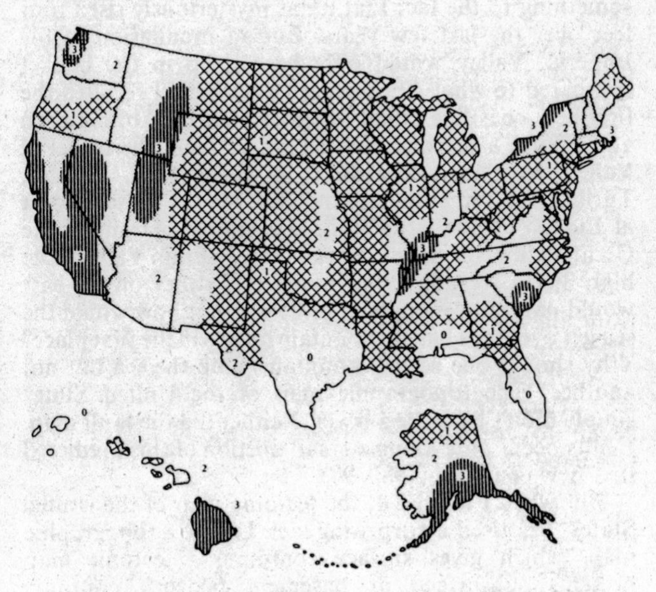

America's Earthquake Vulnerability. This ESSA/Coast and Geodetic Survey map indicates the degree of risk throughout the United States. Zone 0 can expect little or no *damage; Zone 1 can expect* minor *damage (V and VI on the Mercalli scale); Zone 2 can expect* moderate *damage (VII on the Mercalli scale); Zone 3 can expect* major *damage (VIII and above on the Mercalli scale).*

earthquakes—such as those that took most of the lives in the 1964 Alaska quake. But these facts, I reasoned, still couldn't explain how the ocean might move 1,000 miles inland and wash over the western third of the United States. Shades of Atlantis!

From the topographic map of the United States, the nearest reasonable possibility I could imagine was a huge tidal wave washing into the Imperial Valley of southern California, which is several hundred feet below sea level. The Imperial Valley is the area the psychics said would be the first to go. A huge body of water called the Salton Sea lies in this valley. I also imagined there might be something to the fact that it has mysteriously risen four feet over the last few years. But an inundation of the Imperial Valley would only be a drop in the bucket compared to what Abrahamsen described. I recalled the first new coastline (1980–90) indicated by Abrahamsen and Elkins, a seacoast that would go from the Imperial Valley to San Bernardino to Bakersfield, Fresno, Turlock, Sacramento, and then back to the present coast at Eureka. These cities trace the eastern margin of the Great Valley, the Sacramento—San Joaquin Valley. The high Sierra Nevadas that bound this valley on the east would naturally enough stop the sea, but how would the sea get over the coastal mountain ranges in the first place? Why should one set of mountains stop the sea but not another? The topographic map of the United States simply didn't have the answer. Neither did the fault map. Faults occur only along a small portion of this predicted first new coastline (1980–90).

But when I looked at the tectonic map of the United States, I noticed a surprising fact. Unlike a topographic map, which gives surface contours, a tectonic map presents subsurface or basement bedrock contours ("basement" being the solid crustal layer beneath the sedimentary layer)—in other words, *structural* contours. This map showed the major structural features produced by uplift, downwarp, and faulting of the North American strata. It also showed the geological composition of these beds. On this map in the San Joaquin–Sacramento Valley area I saw the symbol for "thick deposits in structurally negative areas." In plain language, this meant that thick deposits of poorly consolidated sedimentary materials underlay the valley, and that the basement rocks of the valley were below sea level. Also, the New York Academy of Sciences recently reported that all is not quiet in the valley. It is estimated that the valley is sinking at the rate of one foot per year and has already sunk more than thirty feet in recent times. With strong enough vibrations from earthquakes, these deposits could give way, just as the heavily faulted deposits lying between the present coast and the valley could give way. And once these deposits give way, the sea could move in over low basement contours.

This is what acutally happened during the 1964 earthquake at Niigata, Japan. The ground shook so that water came bubbling out of the ground. The result was that the poorly consolidated sediments liquefied and turned to quicksand. Something similar happened in the 1906 San Francisco quake. I imagine that if a quake were large enough, anything less than the most solid igneous or metamorphic rocks could be vibrated to the point of failure and liquefaction.

As I pursued this line of reasoning, the tectonic map explained why the sea would stop at the cities named. Unlike the coastal mountains to the west of the Great Valley, the mountains that bounded these cities on the east were underlaid by igneous rock, solid granite. No amount of shaking would make granite give way.

To understand this situation, picture a house foundation built on loose bricks, and another built on a single gigantic brick. The deposits in the San Joaquin–Sacramento Valley would be like loose bricks whereas the granite to the east would be like one solid brick.

The tectonic map also explained why the sea would turn back to the present coast at Eureka: its path would be intercepted north of the valley by solid metamorphic rocks—rocks as strong as granite. (Metamorphic rocks have been subjected to such intense pressure and temperature that they have flowed and fused into the equivalent of gigantic solid brick.)

When I tested this interpretation on the other early

coastal change that the psychics foresaw—the prediction that an arm of the sea would cut across Oregon and stop at Idaho—the results again seemed to hold up. Such an arm would cut across Oregon's heavily faulted beds and be stopped at Idaho because of the Idaho Batholith crossing its path. The Idaho Batholith is a huge, solid granitic mass of the same age as the granitic mass that marks the boundary of the San Joaquin–Sacramento Valley coastline predicted by Abrahamsen.

Excited, I rapidly searched over the tectonic map for a clue to the ultimate coastline foreseen by the psychics—a coastline running through Nebraska, Kansas, and points south. I was stunned to find that the tectonic map explained this prediction also. The western third of the United States is characterized by relatively young, active, and thick geological deposits. They are studded by fairly recent volcanoes and faults. The solid granite zones previously discussed are exceptions because they exist in a very large area of weak deposits. Given a large enough disturbance, these smaller stable areas could be carried off with the rest. The entire area is something akin to a thick but shaky layer cake with a few solid lumps in it. The western deposits, being relatively weak structurally, would be highly susceptible to the tremendous vibration produced by a large-scale earthquake such as the 12.0 magnitude quakes Elkins predicted.

Ignoring topography—that is, forgetting about surface features—I noticed that the tectonic map showed these less stable deposits running right up to the eastern margin of the Rocky Mountains. Cheyenne, Wyoming, and Denver, Colorado, lie just east of the Continental Divide, but Cayce and the psychic group had the sea going much further east than these cities—as far east as North Platte, Nebraska, and the western part of Kansas. Once again the tectonic map supplied a possible explanation. Along the eastern margin of the Rocky Mountains exists a series of underground basins similar to those which presently house the world's oceans. Indeed, the geological record testifies that these particular basins once did house oceans. In the Cheyenne–Denver area, according to the map, there is the immense Denver Basin. Its western edge

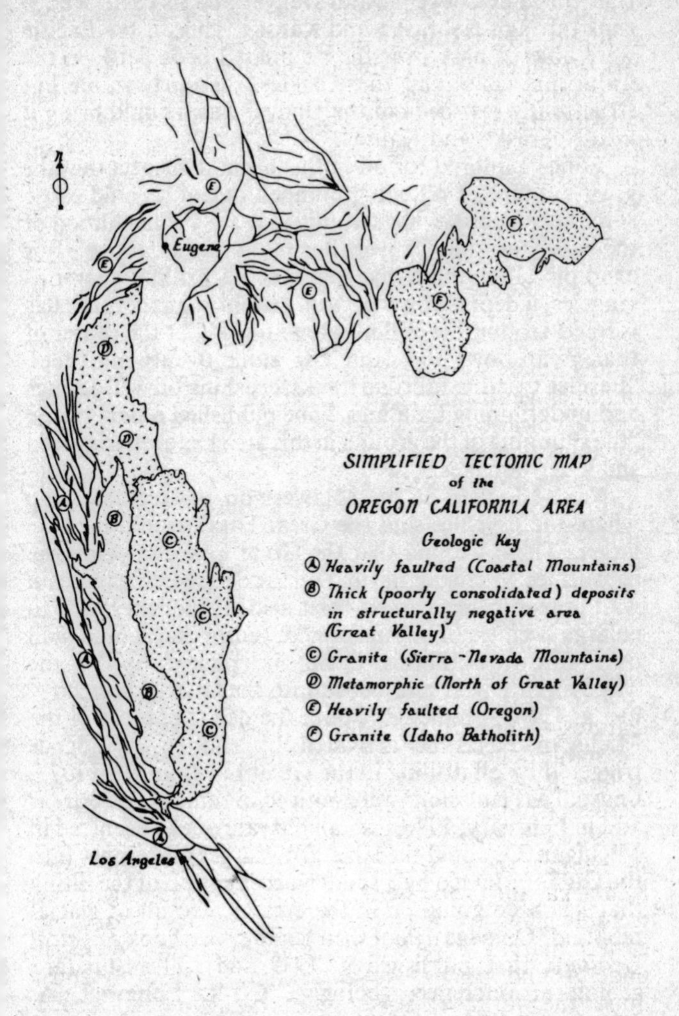

This geological map supports the psychics' projections for the West Coast areas which will undergo destruction during the 1980–1990 period.

runs through Cheyenne and Denver and its eastern edge runs through Nebraska and Kansas. Thus, if the Pacific were able to pass over the decimated beds between the California coast and the Rockies, the fairly stable but structurally low beds of the Denver Basin could bring it into Nebraska and Kansas.

I once explored for oil in this basin. I suspect that the great amount of oil being pumped out of it could cause some subsidence, which would increase the likelihood of the role for the Denver Basin discussed above. For example, after thirty-five years of oil drilling near Los Angeles, a depression measuring twenty square miles has formed around the Wilmington airfield. At the center of this great bowl the land has sunk twenty-nine feet. "Besides twisting railroad tracks, crushing oil well casings and undermining buildings," one published report states, "the slumping of the ground in this area has even triggered small earthquakes."

Next I looked at the Midwest on the tectonic map where the psychics said the Great Lakes would become larger. There I found that the Great Lakes sat atop the subsurface Michigan Basin. Did the deep Michigan Basin wait in readiness to assert itself also? If the lakes were to enlarge because of subsurface contours, they would spill over into the Illinois Basin, be picked up by the Mississippi River, and be led into the Gulf of Mexico—just as Cayce predicted. Could the deep contours of the Michigan Basin be activated by sudden subsidence triggered by oil drilling in the Great Lakes area? In 1973, unusual gas emissions were noticed in some of these areas and in February, 1976, a small but rare quake occurred in Michigan. Or could the lakes grow larger and empty into the Gulf of Mexico by a sudden acceleration of the tilting that has been going on in the area as a result of glacial rebound? On page 318 of their geology textbook *Physical Geology*, first published in 1939 and still a standard geological reference, geologists C. R. Longwell, A. Knopf, and R. F. Flint write:

> An excellent example of tilting on a large scale is offered by the Great Lakes. To the northeast the land

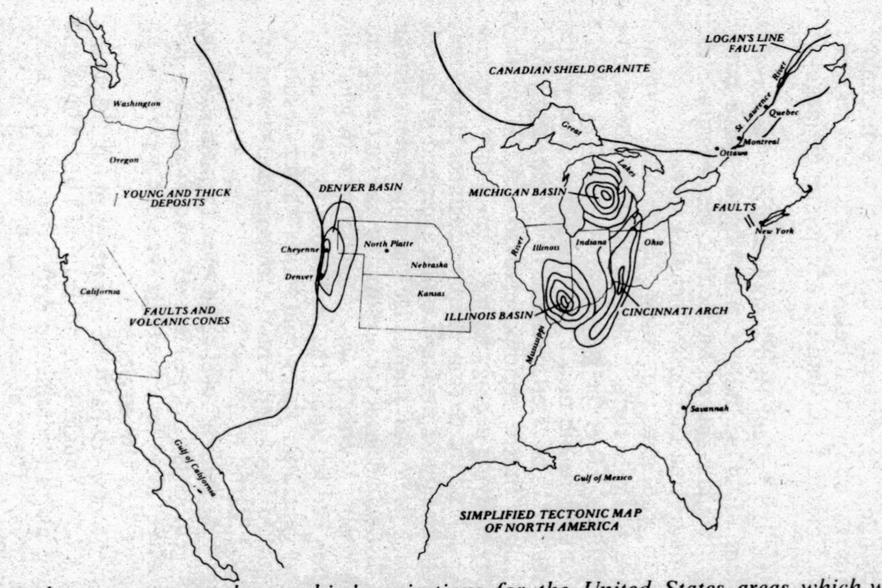

This geological map supports the psychics' projections for the United States areas which will undergo destruction during the 1990–2000 period.

> has risen since the disappearance of the Great Ice Sheet (glacial rebound) and as a result, the lake basins have been tilted southwestward... The tilting movement is still in progress, and has been accurately determined... in 1600 years it could cause the upper Great Lakes to discharge by way of the Chicago River into the Mississippi drainage.

Even the "safety lands" spoken about by Cayce—portions of Ohio, Indiana, and Illinois—came out as being dry and stable on the tectonic map because of the Cincinnati Arch that passed beneath them.

As I continued to study the tectonic map and compared it with the psychics' predictions, I found more correlations. From the geological composition and the contours of the subsurface beds, prediction after prediction seemed to be asserting itself. How could the psychics' predictions fit the tectonic map so well? Were they secretly studying geology books? Even if they somehow had visualized the tectonic map, I remembered how they were in essential agreement with Cayce, and in Cayce's time such geological knowledge was almost nonexistent.

A distinct uneasiness came over me as I sat there in my living room. My mind was catching up with what my intuition had felt about the predictions for some time. It was *my* generation that would witness these catastrophic changes! My discoveries on the tectonic map seemed too wild and fearful to discuss.

But a measure of relief came when I remembered that some geologists, independent of psychic sources, had themselves called for essentially the same great breakup of the western United States.

Enter a new Global Rift! As already noted, the earth's crust is now thought to be made up of a dozen huge and drifting plates ("continental drift"). One description of them says, "These plates constantly interact at their boundaries—bumping, grinding, pulling apart, plunging one beneath the other." This jostling is the cause of many of the world's earthquakes. Two of these plates—the Pacific and the North American—meet along California's infamous San Andreas Fault. Geophysicists believe the

CONTINENTAL DRIFT

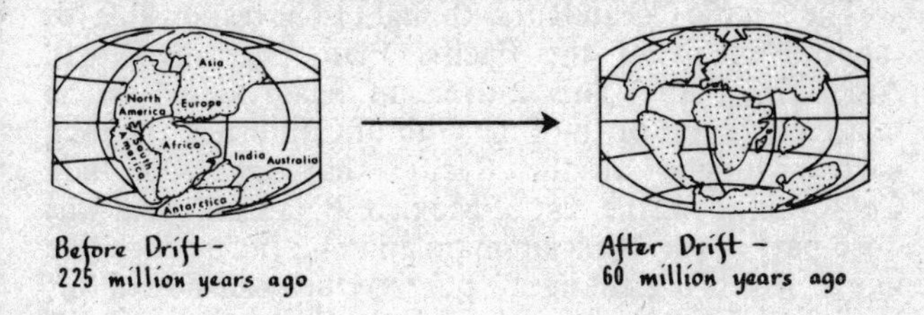

adopted from Robert S. Dietz
and John C. Holden

The movement or drifting of the continents over the ages.

energy for this movement—that is, for continental drift—mainly comes from two mid-oceanic rifts. These are places where the earth's molten interior wells up to solidify, form new crust, and move outward as a result of pressure by the material beneath.

The two primary rifts are in the Pacific and the Atlantic. Along these rifts the sea floor is believed to be spreading, thereby pushing the continental plates ahead of them. In recent years "sea-floor spreading" has been closely observed and studied. Along what is called the Mid-Atlantic Ridge (a ridge bordering the rift) the Atlantic Ocean Basin has been shown to be opening several inches a year. And parts of the Pacific floor, particularly off South America's earthquake-wrenched west coast, have been shown to be opening four times as fast as the Atlantic Ocean Basin. The Mid-Atlantic Rift is held responsible for North America parting company with (once-adjoining) Europe and likewise for South America splitting off from Africa. When the Atlantic Ocean enlarged, the Pacific Ocean correspondingly shrank, its floor being consumed in the wrinkles forming the great trenches around its rim. Thus we have a continuous process of sea floors spreading, plates moving, and the earth's crust being created, rearranged, and destroyed.

The Pacific Ocean Rift is thought to be responsible for the movement of the Pacific Plate past the North American Plate. This movement has resulted in he relatively recent birth of the Gulf of California, the long narrow tongue of the Pacific that separates Baja California from the rest of Mexico. Baja California was once part of the Mexcian mainland and fits its coastline like a piece of a giant jigsaw puzzle. One extension of this spreading-sea-floor-caused rift is the San Andreas Fault system of California. Recently geophysicists, notably Dr. Kenneth L. Cook, chairman of the University of Utah's Department of Geophysics and also director of Utah's seismograph stations, has drawn attention to another extension. After years of studying the frequent earthquake activity at the base of Utah's Wasatch Mountains, Cook concluded that the Wasatch Fault is part of the world rift system—that is, a new rift. Since 1850 more than six hundred earthquakes have occurred in the state along this zone. Dr. Cook thinks the chief mid-ocean ridge in the Pacific—called the East Pacific Rise—extends inland, through the Gulf of California, across Arizona, central Utah, under Salt Lake City, southeastern Idaho, and Yellowstone National Park.

On the North American tectonic map, the path of this ridge or rift is marked by heavy concentrations of parallel faults. In a 1968 interview that appeared in the *University of Utah Review*, Dr. Cook stated his opinion that eventually the United States will be split along this rift line into two continents, separated by the Gulf of California. Dr. Cook is essentially talking about the formation of a new plate. In his view, the Gulf of California rift valley will continue to develop across the surface of these states and a million years from now it will fill with water. The result will be similar to the Red Sea, which separates Africa and the Arabian peninsula. The Red Sea, its floor split by an active rift, is considered by most geologists today to be an ocean just beginning to open.

I marveled once again at how the North American tectonic map stood behind predicted changes and how

Cook's projected coastline was so similar to that projected by the psychics, especially as it cuts through Arizona and passes by Yellowstone Park, which Cook describes as "an intense thermal area similar to a volcanic-type area" in the southwest corner of Montana. (Ominously, in 1969 a large quake struck Yellowstone Park and in December, 1976, three small quakes—4.5, 4.3, and 5.0—surprised residents. In the geological past a portion of the Gulf of California washed over part of California and Arizona.) The psychics' projected coastline cuts farther east in some places than Cook's coastline, the subsurface structural basins seemingly coming into play in their predictions. More important than specific locations, it was eye-opening to see how a professional geophysicist called for the same dramatic earth changes as the psychics.

There were two basic differences between the projections of Cook and the psychics. First, the psychics had all the land west of the new coastline being broken up into islands and/or destroyed whereas with Cook's concept of sea spread, just a narrow sea will form along the rift line. Thus, in Cook's scenario we could expect only the western margin of California to be destroyed—by being consumed or subducted under the Pacific Plate in order to make room for the growing sea. Yet Cook's partial destruction of the western margin of California is still a far cry from the views of many other seismologists, who say that western California will simply drift north past the rest of California on a slow journey toward the Aleutian Islands. According to this view, Los Angeles, for example, will move up to meet San Francisco. Interestingly, this difference in viewpoint between Cook and other seismologists also provides a geological basis for an analogous difference between some of the psychics on this issue. In the previous chapter I noted that Abrahamsen essentially foresaw all the land west of the new coastline destroyed while Elkins, Bernhardt, and Karish foresaw the formation of islands. Cook's "sea-floor spread" would tend to destroy real estate whereas "continental drift" would tend to cut up real estate into islands.

The second and more important difference between the

psychics' and Cook's views was the time scale. Cook said that his coastline would come about gradually, taking more than a million years, while the psychics said that their coastline will come about suddenly, in only one generation.

How can we judge the likelihood of their relative accuracy?

If we go back to the Red Sea example, which Cook himself favors, we observe something very interesting. At the southern end of the rift-formed Red Sea another rift also strikes into the African continent through the Gulf of Aden. This rift region is known as the Afar Triangle. The renowned volcanologist Hanoun Tazieff wrote in *National Geographic* (January, 1973) that the Afar Triangle was very recently covered with seawater. Once-submerged beds have been radiocarbon-dated to just six thousand years ago. Tazieff concludes that "the topography has been created by *violent* events that have occurred in very recent times . . . and are still in progress . . . The triangle seems to be a focal point for new oceans in the making. What is more . . . we can see it taking place."

In this region geologists found volcanic glass that was spewed out and violently cooled under water. They mapped wide-open fissures and fresh faults, explosion craters, and boiling springs. Geologists even saw an earthquake actually shift the crust along a fault line. This is evidence for the psychics' accelerated time scale in ocean formation rather than Cook's slow and gradual change. Ominously, earthquakes frequently take place in the Gulf of California. The adjoining Baja California peninsula seems to shake all the time. A recent five-year study by U.S. seismologists showed that as many as six hundred tiny earthquakes can occur daily on the peninsula, with the average exceeding one hundred daily. Hot spots are common and fresh volcanic cones rise near the gulf's head. For example, a 5.0 magnitude quake took place in the gulf on December 7, 1976. It jolted residents of Yuma, Arizona, and San Diego, California.

The match between the psychics' predictions and the North American tectonic map seemed almost too perfect to be true. Yet it did not stand alone. There were a number of other points to support the predictions of cataclysm. In 1976, the International Geodynamics Project issued a world "heat flow" map, the first of its kind showing the amount of heat flowing outward from the earth's interior. Upon inspecting the listing for the North American continent, we see that the highest heat flow by far was coming from the particular western states covering an area similar to that which the psychics predicted destruction for. Furthermore, in the past, San Diego, Los Angeles, and San Francisco all have been hit by large earthquakes. Major quakes occurred in California in 1857, 1872, and 1906. In the 1857 Los Angeles (Fort Tejon) quake, the west side of the San Andreas Fault moved northward more than thirty feet. In 1872, the convulsion along the Owens Valley Fault brought people into the streets from San Diego in the south to Eureka in the north—literally from one end of the state to the other. If anything, the many California faults are showing more signs of activity in the past few years, as the internal pressure seems to be building up. For example, USGS geologist Robert O. Castle recently discovered that vast areas of southern California, stretching from the Pacific to the Mojave Desert, have uplifted ten inches in the last fifteen years. And while the northern segment of the San Andreas Fault received relief in the 1906 San Francisco quake, the southern section of this fault, near Los Angeles, has been locked and stress has been accumulating since 1857.

In September, 1977, California Institute of Technology scientists began to worry about the sharp increase in the number of small earthquakes along a twenty-mile stretch of the San Andreas Fault near Palmdale, a phenomenon that seems unrelated to the bulging ground around Palmdale. In a ten-month period the scientists recorded at least four hundred microtremors registering between 0 and 3 on the Richter scale. (Quakes this small usually are not felt by people.) Only fifty tremors might be expected in a comparable ten-month period. They say the pattern is similar to the clusters of microtremors now known to have preceded the disastrous southern California earthquake on February 9, 1971. Dr. Don L.

North American locations receiving the highest amount of heat flowing outward from the earth's interior.

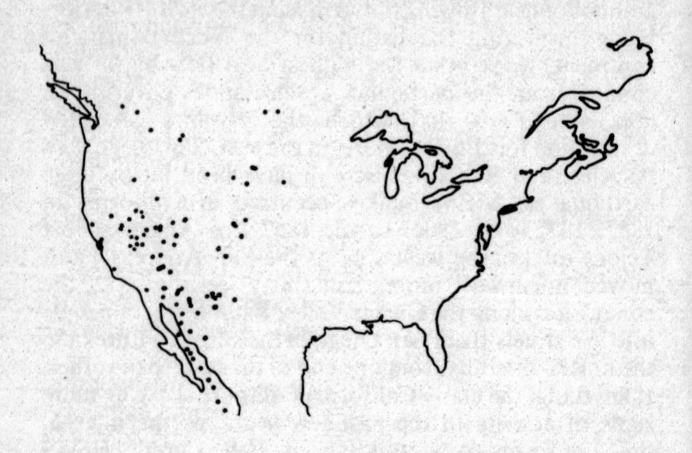

adapted from International Geographic Project (World Data Center for Solid Earth Geophysics)

Note the similarity with the new coastline projected by the psychics for the 1900–2000 period.

Anderson, director of California Institute of Technology's Seismological Laboratory, believes that a moderate to strong earthquake is building up around Palmdale. Another California Institute of Technology scientist, Dr. Karen McNally, says that analysis of other earthquakes that have been preceded by swarms of small tremors has shown that the microshocks have come between two and ten years before the big quake. A similar pattern was found before the disastrous Tangshan earthquake in China in 1976 that took more than 750,000 lives. Ominiously, Los Angeles County highway maintenance crews are finding recurring cracks in the roads in the same 30,000-square-mile area in which the small quakes are occurring.

The prediction of New York City being ravaged by quakes is supported by the fact that several faults are known to cut across Manhattan Island. The towering skyscrapers of Manhattan are not built on solid rock, as most people think. Dr. A.K. Lobeck of Columbia University reported on New York City's faults in *Geomorphology*. When Abrahamsen said that New York City's southern area would be completely lost, he could be referring to the large fault that cuts across the southern part of the island at Fourteenth Street. Consolidated Edison, New York City's electric utility, discovered this chasm when they repaired underground cables. Interestingly, seismologists have recently reported increased seismic activity in New York State itself, a state one usually doesn't associate with earthquake activity, as a result of the Ramapo Fault.

The prediction that Philadelphia and Washington, D.C., will receive some damage brings to mind February 28, 1973, when an earthquake of approximately 4.0 magnitude occurred in seemingly stable Delaware. This led geologists to identify enough tiny quakes beneath Wilmington, Delaware, for them to infer the existence of an active fault, or possibly several faults, in the region. Wilmington is directly between Philadelphia and Washington, D.C.

With regard to the eastern part of the United States undergoing severe disturbances, we know that large earthquakes struck Boston in 1755, New Madrid, Missouri, in 1811, and Charleston, South Carolina, in 1886. The 1811 New Madrid quake in southeastern Missouri was the single most powerful earthquake in American history. It rattled buildings as far away as Chicago, Washington, D.C., and New Orleans. Large areas of the Missouri floodplain also sank far below their previous levels. One resident of the then sparsely populated area wrote: "The whole land was moved and waved like the waves of the sea. With the explosion and bursting of the ground, large fissures were formed, some of which closed immediately, while others were of varying widths, as much as 30 feet." Recent small and perhaps premonitory quakes have been taking place within a few hundred miles of New Madrid. For example, in 1975 and 1976 quakes broke windows and knocked bricks from chimneys and dishes from shelves several times in Tennessee. Dr. Robert M. Hamilton (mentioned in the first chapter) indicated that he "would not be surprised if there were a destructive earthquake" in the Mississippi Valley area. Note that today the populous city of St. Louis lies only 150 miles to the north of the site of this 1811 quake, and that Memphis is about the same distance to the south. Government geologists, worried about the earthquake hazard to proliferating nuclear power plants, have recently begun to study the thousands of old fault lines lying deep beneath the eastern states.

Geologists admit that they neither know what caused the deep and still ill-defined faults that crisscross the eastern United States, nor what could now activate them. In 1976, the USGS began installing a network of eighty seismic stations in the Northeast in order to estimate the earthquake hazard there. In the January, 1973, issue of *Geotimes*, Dr. Kark Kisslinger of the University of Colorado (one of the United States' seismic centers) concluded, "Although the number of large earthquakes east of the Rocky Mountains is much smaller than to the west, the much larger area of high intensity for a given magnitude in the east makes the long term risk, in terms of potential damage to property and loss of lives, roughly as great as in the west." Dr. Kisslinger based his conclusions on the virtual absence of large earthquakes in the eastern United States for many years, and the pattern of premonitory earthquakes that is now appearing in the Northeast, Southeast, and parts of the Midwest. Some believe that the properties of the earth's crust in the East substantially compound the risk of massive damage resulting from a major quake, since the earth's crust in the East is relatively rigid, which would transmit the destructive energy released in a quake over a much wider area. This contrasts with the more highly faulted structure of the West, which relatively quickly attentuates an earthquake's shocks.

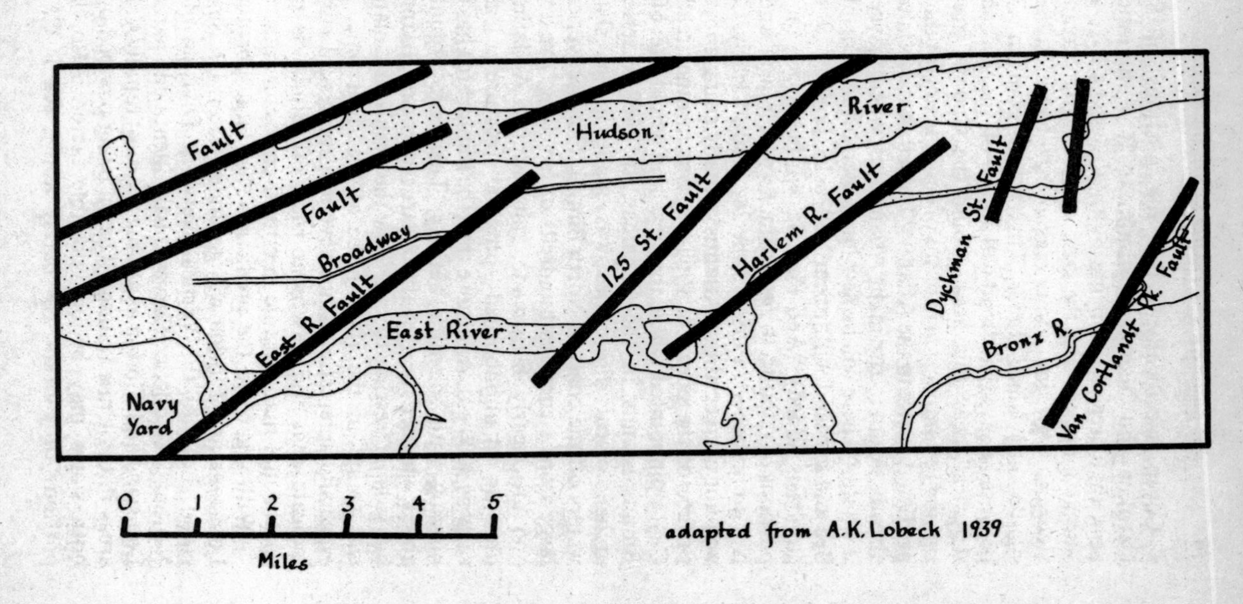
Major Faults Running Through New York City
Fault
Fault
Hudson
River
Broadway
East R. Fault
125 St. Fault
Harlem R. Fault
Dyckman St. Fault
Van Cortlandt Pk. Fault
Bronx R.
East River
Navy Yard
0
1
2
3
4
5
Miles
adapted from A.K. Lobeck 1939

Cayce specifically called for portions of South Carolina and Georgia to disappear. It is interesting to note that precise surveying by the Coast and Geodetic Survey indicates that the land surface in the Savannah, Georgia, area for some reason has begun subsiding significantly since 1933. This might confirm Cayce because he says that "land will appear off the east coast of America" near this area. Could the Georgia area be sinking to compensate in terms of equilibrium for the land predicted to rise in the Atlantic? Indeed, it seems land has already begun to rise in the Atlantic just where Cayce said it would. "There will be new lands seen off the Caribbean Sea and *dry* land will appear" (3976-15, January 19, 1934, emphasis added). "And Poseida will be among the first portions of Atlantis to rise again. Expect it in '68 and '69. Not so far away!" (958-3, June 28, 1940). Clearly, Cayce was not correct about Atlantis actually rising above the surface of the ocean in 1968–69. But in other readings Cayce pinpointed where this new land would rise: off the Bimini Islands in the Bahamas. In 1968 a series of archaeological discoveries was made in the shallow waters off the coast of North Bimini. The discoveries are large stone structures hundreds of feet long, like a wall, and are clearly man-made, not natural. A dating of one—as yet unpublished—puts the structure at least as early as 3000 B.C. And since then the Bimini area has been buzzing with scientific activity. There are arguments about whether the megalithic structures are Atlantean, but it is interesting to note how the prehistoric material was found in the right place at the right time. It is especially interesting since several geophysical surveys indicate that the basement rock off Bimini is more continental than oceanic in crustal characteristics.

With regard to the prediction that large portions of Louisiana and Florida will go under water, Dr. Samuel Hand, head of the Vertical Network Division of the National Geodetic Survey, has identified parts of Louisiana and Florida as among the most rapidly sinking areas in the United States. The oil being pumped out of both states may be one possible factor. Also, large portions of Florida are barely above sea level; any significant rise in sea level would quickly cover much of the state. The structural contours of the tectonic map show coastal Louisiana and Florida to be among the lowest areas in the United States, some places as much as 22,000 feet below sea level.

The psychics also specifically called for land to rise in the Bering Strait area before the turn of the century. Land rising there would form a bridge between Asia and North America. Archaeologists believe the ancestors of the American Indians used just such a bridge when they first entered the Americas. In fact, this bridge has risen and fallen several times in the past, so geologists would essentially agree with the psychics in expecting it to rise again. Land last emerged twenty-five thousand years ago, only to slide back into the waters ten thousand years ago.

The prediction that the nearby Aleutian Islands will disappear as the Bering Bridge rises would not surprise those who know that as a result of the 1964 Anchorage, Alaska, earthquake an estimated forty thousand square miles of land dropped as much as eight feet and about twenty-five thousand square miles of land were raised as much as thirty-three feet. The theory of sea-floor spreading, with land on the edges of the sea floor being consumed in ocean trenches to make room for new land being added to the sea floor, will more than accommodate this prediction, except for the time scale. The Aleutian Islands mark the border of an ocean trench. To make matters worse, the islands consist of a chain of volcanoes, some still active. They lie in the ocean like a string of smoldering firecrackers just waiting to be touched off.

The Japan Current now swings by the Aleutian Islands and Alaska on its way down to the California coast. The prediction that it will shift to a more southerly position seems to fit well in the general scheme in which the Aleutian Islands, Alaska, and California all undergo major disturbances.

Although the United States is supposed to undergo vast changes, Canada is seen as being fairly stable, with certain exceptions. Vancouver will be inundated, according to the predictions. The tectonic map shows that this is quite possible since Vancouver is sitting atop thick, poorly consolidated deposits in an area with a deep depression (or "syncline") in its basement—the same configuration underlying the San Joaquin–Sacramento Valley.

The tectonic map also testifies to Canada's projected stability because much of the basement rock underlying Canada is solid granite. Geologists call the extremely stable areas of Precambrian rock (the world's oldest) that rise above the ocean "continental nuclei" or "shield areas." Each of the major continents has a shield area on and around which younger rocks were deposited. North America's is the huge Canadian Shield, which blankets the eastern two-thirds of Canada.

Despite this apparent stability, the psychics foresaw the southeastern cities—Quebec, Montreal, and Ottawa—being damaged by quakes, the St. Lawrence River being enlarged as a result of faulting, and the St. Lawrence Seaway enlarged and thus severely damaged. A close look at the North American tectonic map once again supplies supporting evidence. Violating the integrity of the shield in this area is the Logan's Line Fault. This long fault runs down the middle of the St. Lawrence River and moves landward to pass under Montreal and Quebec, which border the river. One arm of the fault even cuts westward to pass under Ottawa.

I marveled at how precisely the psychics' predictions for Canada worked out on the tectonic map. But there was more. The St. Lawrence River Valley also has a historical record that confirms the presence of a local source of earthquakes capable of producing heavy destruction. In 1663 a major quake struck the then sparsely populated area and was felt in all of eastern Canada and the northeastern United States. In 1884 a moderate-sized shock occurred, and recently, on June 15, 1973, a minor quake (4.5 on the Richter scale) rattled the area. Dr. Benjamin Howel of Pennsylvania State University indicated that the shock may very well have been the precursor of a more intensive series of earthquakes centered in the St. Lawrence River Valley. Indeed, seismologists such as Richter have long identified this area as one of high risk.

Furthermore, the sequence of events that is to precede the total destruction of California and the Decade of Cataclysm is supported by the tectonic maps, historical precedent, and current trends. In this sequence, devastation starts in India, goes to Japan, then to Italy, and finally to Martinique before it reaches the United States. First, earthquakes and tidal waves are supposed to hit India, especially its east coast. Earthquakes can indeed generate tidal waves, and of immense proportions. For example, in August, 1976, tidal waves over fifty feet high engulfed the white sand beaches of Mindanao Island in the Philippines, sweeping away survivors of the quake that had struck a day earlier. The largest recorded seismic sea wave, or *tsunami*, was 220 feet high. It appeared off southwestern Alaska after the 1964 Good Friday earthquake. Just before this quake the water disappeared from coastal bays, only to come roaring back in several successive tidal waves after the quake struck. These walls of water were much smaller than the 220-foot-high wave that appeared later off the coast; nevertheless these smaller waves wiped out villages, flooded towns, and killed people as far away as Oregon. The 131-ton crab boat *Selief* was swept in and out of Kodiak harbor on three massive waves, finally coming to rest in back of the Kodiak schoolhouse five blocks from shore. Seismic sea waves have been observed to travel as fast as 490 miles per hour and as far as 9,000 miles. Tidal waves can be every bit as destructive as earthquakes.

Large earthquakes and tidal waves are nothing new to India. In 1937 one of the worst earthquakes of all time took three hundred thousand lives in Calcutta. In 1905, twenty thousand lives were lost in an earthquake. Another sixty thousand lives were lost in 1935, and in 1950, thirty thousand more. On December 28, 1974, an earthquake killed more than five thousand in nearby Pakistan. Although India has experienced many typhoons, only a few tidal waves have struck the country. Moreover, most of these tidal waves have struck only in the last few years. On November 13, 1970, a cyclone-driven tidal wave from the Bay of Bengal killed more than two hundred thousand. And on September 29, 1971, eastern India was hit by another cyclone and tidal wave from the Bay of Bengal that killed ten thousand people. Might the psychics' predictions have anticipated some new pattern developing for India? Western and northern India are active tectonic areas, at the edge of the Indian crustal plate that has slammed into Asia and pushed up the towering Himalayas in the process.

In the psychics' key warning sequence, after India these tidal waves and earthquakes are to hit Japan and eventually cause large parts of it to go under the sea. In fact, large parts of Japan have already sunk beneath the sea. The December 24, 1977, issue of *Science News* reported that evidence obtained by deep-sea drilling shows how present-day Japan "is but a small part of a former landmass that, many millions of years ago, extended about 300 kilometers farther into the Pacific (toward what is now a deep undersea valley known as the Japan Trench)." The scientists who conducted the study reported in *Science News* said that the sedimentary sequence from the core samples revealed that the "ancient landmass subsided rapidly from above sea level (about 25 million years ago) to more than 2 kilometers below the ocean surface." Today Japan is still in an active tectonic area. It lies at the juncture of two massive crustal plates, one of which is being thrust under the other rather than sliding by, which is what usually happens. Japan also lies along the geologically active Circumpacific belt, which marks the rim of the Pacific Ocean. This is the "Ring of Fire" along which 80 percent of the world's earthquakes take place. Japan is also crisscrossed by major active fault systems. All these facts make Japan one of the most geologically unstable areas of the world. The earthquake hazard in Japan is even greater than that of California. It is a toss-up which locale—Japan or California—will undergo a major quake first. And Japan is much more densely populated than California.

Japan already has a record of disasters that illustrates its geological instability. In 1896 an earthquake-generated tidal wave killed twenty-seven thousand people. In 1923 an earthquake that also produced a tidal wave destroyed one-third of Tokyo and most of Yokohama, taking the lives of one hundred and forty-three thousand in the process. Vertical displacement has been measured as one plate is thrust under the other. In some cases the surface subsidence rate has been calculated as several inches annually. One geologist concluded that the subsidence along the coast of Toyama Bay, which caused the submergence of an entire forest, was due to movement of a block of the crust. Slow and uniform sinking is no surprise to Japan, but an illustration of sudden and catastrophic sinking may be found in the fact that in the 1923 Tokyo earthquake in adjoining Sagami Bay the sea bottom in one area sank 1,310 feet. In 1933 one of the strongest earthquakes ever recorded, 8.6 on the Richter scale, occurred 100 miles off Japan's Honshu coast. On June 17, 1973, a quake of magnitude 7.9 struck sparsely populated northern Japan. And in July, 1973, Japanese seismologists detected unusual earth movement and faults in the Tokyo area.

This new and ominous seismic activity is a precursor of things to come: seismologists are nearly unanimous that Japan is due for a cataclysmic earthquake. This has led the Earthquake Research Institute of Japan to make a clear statement about national research priorities:

> All through her history, Japan has suffered frequent great earthquakes, and each of these has caused a large number of casualties and an enormous amount of damage. *It is certain* that earthquakes will occur in the future in a similar way as in the past and disasters as caused by them must be prevented as far as possible by ourselves. Prediction of earthquakes is *an urgent necessity* of the nation and is also the final aim of scientific endeavors in the field of *seismology in this country*. Seismological studies up to the present do suggest possibilities of realizing this aim of earthquake prediction. In order to make such possibilities a practical reality, deep understanding and ample financial support by the government is indispensable, in addition to the

> constant and conscientious endeavors of all the researchers concerned [emphasis added].

The Japanese government now fully endorses and financially supports earthquake research activity. It is the first country in a high-seismic-risk area with a clear public policy regarding earthquakes. Officials actively conduct earthquake drills and have developed earthquake response plans. Monitoring systems are being installed across the country. And judging from the number of best sellers on this topic in Japan and the earthquake detention kits sold in department stores there, the people also clearly recognize the possibility of cataclysm.

After India and Japan are hit, the final signals in the key warning sequence will be the eruptions of Mount Vesuvius in Italy and Mount Pelée in Martinique. The eruption of Vesuvius is to be triggered by a violent earthquake and these catastrophes are supposed to cause damage as far away as France and Scandinavia. Two things should be noted here. First, the association between earthquakes and volcanic activity is fact. Richter has pointed to the close correlation of volcanic lines with the epicenters of intermediate-depth earthquakes. The 6.9 quake and aftershocks that struck Italy on May 6–8, 1976—its worst quake since 1915—resulted from the convergence of plates in that area. (The crust on which Europe rests is moving eastward while another plate [the Turkish Plate] goes westward.) Second, regarding the possibility of an Italian earthquake–volcanic eruption causing damage to distant countries, it should be noted that the May, 1976, Italian quake did minor damage in Yugoslavia, France, Belgium, Germany, and Austria, sending many people fleeing into the streets in panic. And the 1883 explosive eruption of Krakatoa, in Indonesia, illustrates the far-reaching effects of a major eruption. Geology professors L. Leet and S. Judson relate in *Physical Geology* that

> the noise of the eruption was heard on Rodriguez

> Island 3,000 miles away across the Indian Ocean and a wave of pressure in the air was recorded by barography around the world . . . Columns of ash and pumice roared miles into the air, and fine dust rose to such heights that it was distributed around the globe and took more than two years to fall. During this time sunsets were abnormally colored all over the world.

Seismic sea waves hit land as far away as Cape Horn and England. Rocks were thrown to a height of thirty-four miles and dust fell ten days later at a distance of 3,313 miles. This explosion has been estimated to have had twenty-six times the power of the greatest H-bomb ever detonated, and yet it was probably only about one-fifth the size of the volcanic explosion in 1470 B.C. of Santorini, a volcanic island in the Aegean Sea.

Mount Vesuvius has a long record of eruption. The worst and most well known was in 79 A.D. when the mighty giant destroyed and buried Pompeii and Herculaneum. In 1631 it destroyed Torre Annunziata, Torre del Greco, Resina, and Portici. The most recent major eruption was in 1944. There is an air of foreboding when we realize that Vesuvius hasn't erupted in well over thirty years—especially when we consider the worldwide increase of volcanic activity since 1955.

The question now is this: is Mount Vesuvius ready to erupt again and if so, when? In a 1969 London *Daily Express* article, Professor Imbo, director of the observatory built on the side of Vesuvius, was quoted as saying that the next eruption of Vesuvius "could be a matter of years or it could be a matter of weeks . . . But we now have certain evidence that Vesuvius is ready for eruption once more." According to Abrahamsen, the eruption will be in the late 1980s. What are we to make of the recent eruptions of volcanoes relatively near Vesuvius? Mount Etna and a volcano on Stromboli Island—both only 185 miles from Vesuvius—have shown activity recently. Etna erupted violently in 1964, 1974, 1976, and 1977, while the volcano on Stromboli erupted in 1967. Alfred Webre and

RECENT VOLCANIC ERUPTIONS OF
VESUVIUS (ITALY) AND PELÉE (MARTINIQUE)*

Vesuvius	*Pelée*
1779	
....	1792
1794	
1850	
....	1851
1858	
1900	
....	1902
1903	
1926	
1929	
....	1929–32
1944	

* Data from *Encyclopaedia Britannica*, 1972.

Phillip Liss note in their book *The Age of Cataclysm* that these two volcanoes along with Vesuvius are located along the same general Mediterranean fault system and are presumably responsive to the same set of earth forces.

Since the psychics indicated that the eruption of Vesuvius will trigger an eruption of Mount Pelée, we should ask whether Pelée is in fact showing such signs. The geological basis for such a possible connection will be examined in a later chapter. For now, however, let's glance at the more recent eruption records of Vesuvius and Pelée. They are quite revealing.

While these two volcanoes are quite far apart, their eruption records show a remarkable correlation. Shortly after Vesuvius has fired up in the past, Pelée has often kicked its heels. Eruptions of Pelée are something to be reckoned with because when it turned active in 1902 it killed all but two of the capital city's forty thousand inhabitants as it belched out a poisonous sulfurous gas.

We get an indication of Pelée's readiness by the 1972 eruption of Soufrière, a volcano on the island of St. Vincent, only seventy miles away from Pelée. Soufrière had last erupted in 1902, the year of Pelée's last eruption. It would be a fair guess to say that these very close volcanoes respond to the same inner earth processes in that area. In fact, the West Indies island chain of which they are a part is predominantly a volcanic chain of islands much like the Aleutians. Thus, a certain uneasiness came over me when on August 17, 1976, French officials evacuated seventy-two thousand people living near the volcano on Guadeloupe Island a week before it erupted. Guadeloupe is part of this West Indies chain and sits just ninety miles north of Martinique. When this volcano, called La Soufrière (the name is similar to St. Vincent's Soufrière), did erupt, it didn't do so as powerfully as expected, and scientists think that it will be heard from again soon. No plans have been made to move the refugees back to their homes. Volcanoes are natural outlets for molten rock from far below the earth's surface. If the pressure is not again released at La Soufrière, perhaps Pelée will stand in for it in the late 1980s, only a decade away.

As if all this activity were not enough, in 1976 the planet began to exhibit the pattern of sequential activity that Cayce and the other psychics pointed to as warning signals for the greater destruction to come. In this sequence, we'll designate the Indian earthquake and tidal wave as (A). Leaving this aside for the moment, we then have disturbances in Japan (B), Italy (C), the West Indies (D), and after a period of time the west coast of North America (E). On August 17, 1976, Honshu, Japan (B) was hit by a moderate quake. On August 17, 1976, an earthquake also hit fifty miles north of the Italian (C) city of Naples, just outside the Vesuvius area. And on August 21, 1976, the volcano on Guadeloupe Island in the West Indies (D) erupted. A day later a fairly strong quake (6.0 on the Richter scale) occurred off the Kenai Peninsula in

Alaska on the west coast of North America (E). Except for India, we have the sequence the psychics detailed. These events weren't conveniently selected by me from many hundreds of such events that occurred in 1976. Rather, they came from a newspaper compilation of the major catastrophic events of 1976, a list of only twenty-two events that also included other natural disasters such as floods, droughts, and hurricanes.

Am I saying the sequence has begun? No—at least not according to the consensus of my psychic informants. The Indian events must be there first. But the *pattern* shown in 1976 is interesting and, to me, significant.

As if to emphasize its importance, 1976 also saw some other events that brought the predictions of Cayce and the psychic group to mind. First, an occurrence that, while not as precise as the above sequence, is nevertheless provocative. Recall that Cayce said "the upper portion of Europe will be changed in the twinkling of an eye" (3976-15, January 19, 1934). The year 1976 was ushered in by a severe storm hitting northern Europe, killing at least thirty-eight persons, destroying crops, disrupting shipping, and causing floods in the Netherlands, Denmark, and Germany. At the same time Etna erupted with what one account called "unusual violence." And on the same day that Etna erupted, January 2, 1976, Los Angeles was struck by a magnitude 4.2 quake. That day also Greece and the New Hebrides Islands were struck by quakes.

This leads me to suspect that Cayce's prediction, made in 1932, was based on his perception of a hitherto unrecognized pattern of functionally connected events stemming from unknown processes deep in the earth. In other words, the earthquake pattern (B-C-D-E) moving from Japan to Italy, the West Indies, and then to the west coast of America didn't occur in 1976 by accident. On April 9, 1932, Cayce was asked, "How soon will the changes in the earth's crust activity begin to be apparent?" Cayce's answer, as related earlier (see page 12), seems to indicate when the official countdown to the future destructive activities he foresaw would begin, a countdown that begins before the key warning sequence. His answer is internally consistent because the South Pacific is, in fact, almost opposite Mount Etna. To be precise, Etna is at approximately 15 degrees east longitude. The area opposite on the other side of the globe is at 165 degrees west longitude, approximately the location of the New Hebrides Islands. And on January 2, 1976, the New Hebrides were struck by a strong quake (6.8 on the Richter scale), while on the same day Etna erupted. January 2, 1976, was the day the simultaneous events Cayce foresaw for Etna and the South Pacific indeed came to pass! *According to Cayce, the countdown has now begun.*

After evaluating the dramatic North American changes and the key warning sequence the psychics predicted, it seemed almost anticlimactic to review their predictions for the rest of the world. It quickly became apparent that these predictions, like those of North America, also fit with tectonic information. The areas they singled out for cataclysmic earthquakes were all in tectonically active areas—fault zones, volcanic areas, and plate boundaries.

The inundations called for by the psychics could presumably occur as a result of earthquake-generated tidal waves and high tides, as well as from general subsidence. For example, in the 1964 Alaska quake, tides rose eight feet higher than usual along coasts almost three thousand miles away. A slightly higher tide rise would have water breaking, for example, over the dikes of northern Europe. In Venice, where the sea already claims a new piece of land each day, such a high tide would be devastating. Finally, the new land predicted by the psychics would rise as a result of new volcanoes emerging, and land lifting as a result of glacial rebound. A condition of "overbound" might even occur where ground that has risen too fast and/or too much since the weight of the glaciers that covered Europe ten thousand years ago has been removed might one day suddenly drop back to compensate—a condition akin to a rubber band snapping back after having been stretched.

Since quakes are rare in northern Europe, the huge

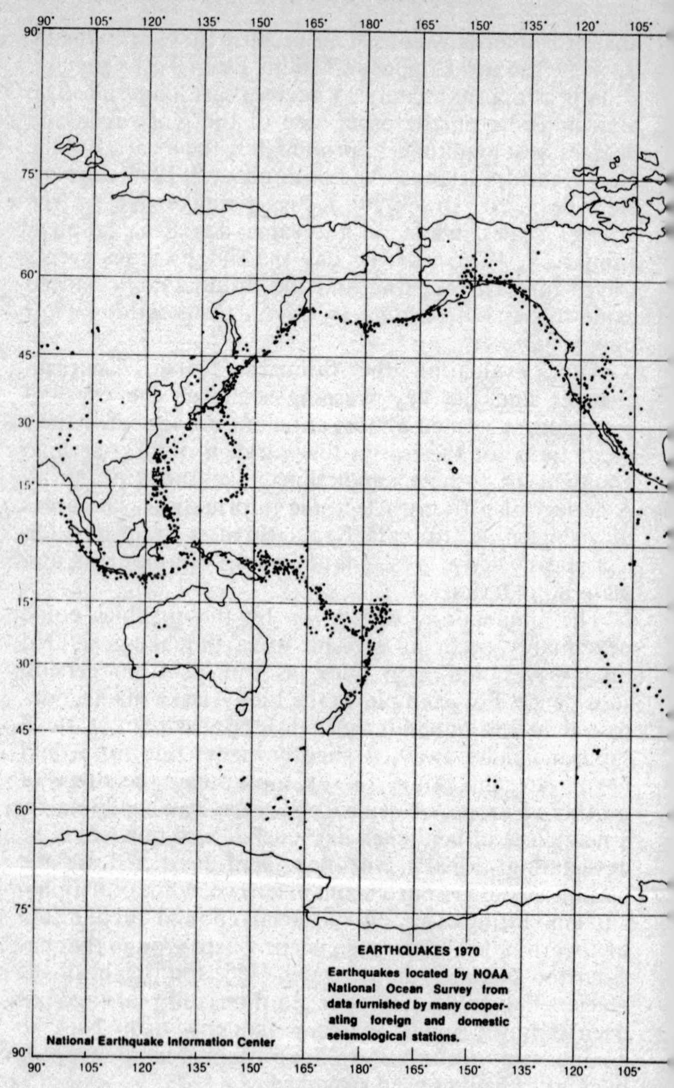

Worldwide earthquake belts indicating locations of past earthquakes and future trouble spots.

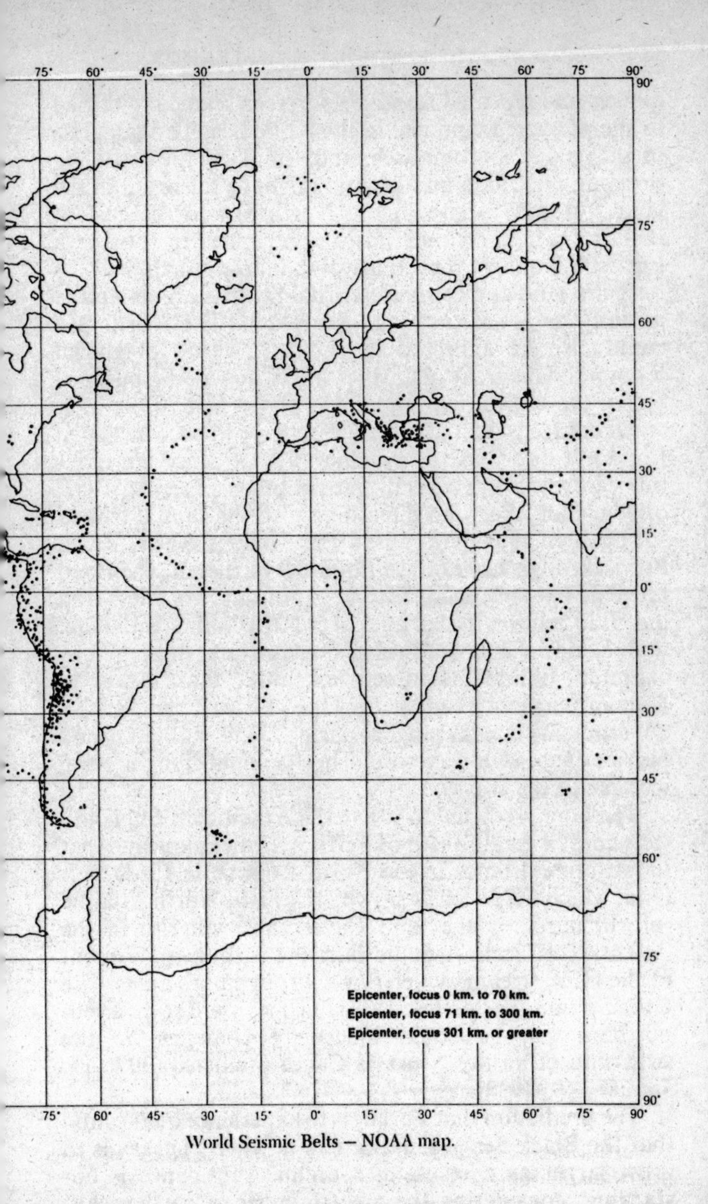

World Seismic Belts — NOAA map.

quakes and repeated floodings foreseen seemed to be tied to the predicted land rise in the Baltic Sea. This land rise might be due to glacial rebound. Similarly the predicted flooding of Britain and Ireland might be the result of the land foreseen to rise in the North Sea to the east of Britain, which would also contribute to northern Europe's troubles. The relatively shallow North Sea was once dry land and connected Britain with the mainland as recently as seven thousand years ago. The North Sea could also be expected to rise as a result of glacial rebound. The great quantities of oil now being pumped out of the North Sea might also play a role.

Of course it won't take much to have the ocean "lap at London's door" as the psychics predict, since ten times this century it has already nearly taken place. In 1953, only a change in wind direction saved the capital from a spring high tide that caused the Thames River, which flows through London, to flood. Nevertheless, the flood killed 307 people, made fifty-two thousand homeless, and did $120 million in damage. More foreboding is the fact that London and southeast England are sinking. For example, London is fifteen feet lower than it was in Roman times. Recognizing the threat, the British government is planning a daring $750 million flood barrier that some say would be the eighth engineering wonder of the world.

The land predicted to rise in the Atlantic off England's west coast is reminiscent of Surtsey, a volcanic island that burst into existence in the North Atlantic in 1963. This land, which the psychics say will rise in the North Atlantic (off Britain), or the land Cayce says will rise in the Atlantic (off Bimini), could interfere with the movement of the Gulf Stream, which flows through both areas. The Gulf Stream warms northern Europe; were it diverted, northern Europe could indeed be changed "in the twinkling of an eye," just as Cayce predicted (3976-15, January 19, 1934).

The prediction that Turkey is to experience quakes and that the Black Sea which borders northern Turkey is to grow larger as a result of shoreline subsidence is not startling, considering the great number of earthquakes that have hit Turkey in the past. The quake that struck northern Turkey in 1939 took thirty thousand lives. The 7.6 quake that struck Turkey in November, 1976, was the worst it has experienced since then. Houses crumbled throughout an area of three hundred square miles. More quakes can be expected in the area because of the Anatolia Fault, part of which runs along the Black Sea coast.

The prediction of a quake in Greece brings to mind the quake and more than forty after tremors that struck Greece on January 2, 1976. I've already discussed the prediction of an Italian quake in the section dealing with the key warning sequence, but to the May 6, 1976, Italian quake should be added the Sicilian quakes of 1908 (which took seventy-five thousand lives) and of 1915 (thirty thousand lives).

The prediction of a large quake for Yugoslavia brings to mind the highly destructive 7.3 quake that hit adjoining Rumania on March 6, 1977, and produced a state of nationwide emergency. In fact, Yugoslavia also reported deaths as a result of this quake, and in Naples, 1,400 miles away, glassware rattled.

Some of the psychics predicted that the Mediterranean Sea would be closed by land rising between Spain and Morocco. The narrow Strait of Gibraltar separates the two countries and provides the Mediterranean Sea with its only natural outlet. The geological record shows that the Mediterranean Sea has been closed at least once already by land rising in the Strait of Gibraltar. And the 3,300-foot sea-floor rise off the coast of Morocco caused by the 1960 Agadir quake provides food for thought.

The earthquakes predicted for the Middle East come as no great surprise, considering that this area is almost totally surrounded by active tectonic zones. On the west there is the active rift zone that runs down the middle of the Red Sea. On the south there is the active rift zone that runs through the Gulf of Aden south of Saudi Arabia. On the north there is the plate boundary that runs across northern Iran. And Iran already has quite an earthquake record. Since 1962 nearly thirty thousand people have lost their lives in three major quakes.

Most of the psychic predictions, you recall, were made in 1972–73. The recent quakes in China already partially fulfill their statements of severe quakes for China, especially in the Peking area. In 1975 the Mukden quake (7.3 on the Richter scale) took many thousands of lives (there is no official report of the number involved). The 1976 Peking quake (8.2) and its severe aftershock (7.9) killed over seven hundred thousand people and injured nearly eight hundred thousand, according to reports in the *South China Morning Post*. But the psychics seemed to have even bigger things in mind. And the experience from these quakes could explain why they said China will recover fairly quickly from the colossal quakes foreseen to strike in the Decade of Cataclysm (1990–2000). The 1920 Kansu quake took one hundred eighty thousand lives and the 1932 Kansu quake took seventy-two thousand lives but Abrahamsen seems to have in mind something like the 1556 Shensi quake, which left almost one million dead. Today such a quake could take several million lives. Thus in a matter of moments a quake could kill more people in China than the total number of deaths our country has sustained in wars from Revolutionary times to the present. The psychics also foresaw lakes being formed west of Peking. The Sang-kan and other rivers that flow west of Peking are likely candidates to be dammed into lakes by land shifts brought about by earthquake activity.

With regard to the southern tip of South America being cut away and the continent's west coast being inundated, its position along the geologically active "Ring of Fire" might provide an explanation. The Pacific Plate runs along South America's coast and cuts across its southern tip. Much of the tip is already broken up into islands and both parts of South America have long been plagued by earthquakes. For example, a 1937 quake in Chile took thirty thousand lives and a 1970 quake in Peru took sixty-seven thousand. The land predicted to rise off the west coast of South America might be pitched up by sea-floor quakes or it might rise like the volcanic island that appeared off Ecuador in 1960. But since the psychics said this land was once inhabited and represents the lost continent of Lemuria, where an as-yet-unrecognized civilization supposedly flourished, a faulting up of the sea floor would be more likely. If they are correct, block faulting, i.e., large-scale faulting along the Pacific Ocean's active rim, was probably how this island was lost in the first place.

In Indonesia, mighty Krakatoa is to erupt again. The tidal waves associated with its 1883 eruption took more than thirty-six thousand lives. Indonesia likewise lies along the Ring of Fire; in addition we have the effects of the many millions of barrels of oil being removed from the ground in this area.

Hawaii is supposed to slip under the sea. The tectonic map made me think of the phrase "ashes to ashes, dust to dust" because the island chain was born of violent volcanic activity. This 1,500-mile volcanic island archipelago represents only the tops of a much longer underwater volcanic chain—2,200 miles of submerged volcanoes—which the psychic group, with the exception of Bernhardt, says Hawaii will join. Geologists believe that this entire 3,700-mile chain was built by a crustal plate progressively sliding over a stationary hot spot deep within the earth. A kink or sharp turn in this chain indicates that the drifting Pacific Plate once abruptly changed course.

The "big island" of Hawaii, with Mauna Loa, the world's largest volcano, and the many volcanoes in its Volcano National Park, is at the most active end of this chain. In 1951 a strong quake and tidal wave hit Hawaii, and in 1973 a 6.2 quake centered on Hawaii caused landslides and some buildings to collapse. Throughout 1970, the volcano Kilauea on the island of Kauai was unusually active. Is this recent activity a warning for the near-total destruction the psychic group foresaw for the islands? In 1972 the President's Office of Emergency Preparedness concluded, "The people, property, economy and ecology of the area surrounding the active volcanoes in Hawaii . . . are endangered by the threat of future volcanic activity." Indeed, as I was completing this book, the *Los Angeles Times* on March 15, 1977, warned that "one of the biggest eruptions of the century is about to occur..." at Mauna Loa. Volcanologist John P. Lockwood told the *Times* that

> the mountain is swelling like mad. It's pulling apart, stretching out . . . the whole inside of the mountain has been shaking without stop for months, rocked by hundreds, often thousand of earthquakes a day. It's priming for a massive eruption . . . Activity within the mountain is incredible . . . survey lines show that in the last six months the whole top of the mountain has expanded six inches.

After considering the Hawaiian volcanoes, I turned to the next of the psychics' predictions—the ice of the Arctic, Antarctic, and Greenland. These are to undergo great change also. For example, Cayce said, "There would be upheavals in the Arctic and the Antarctic" (3926-15, January 19, 1934). The quite unusual volcanic activity during the past five years in the Antarctic supports this prediction. Earthquake activity is quite rare there, which makes the February 8, 1971, earthquake noteworthy.

The most fearsome of the psychics' predictions is the pole shift. Distressingly, there is substantial evidence for the earth flipping on its rotational axis, with resultant worldwide climatic changes and almost unimaginable destruction.

As I pondered how time would soon attest to the accuracy of the vast array of changes the psychics called for, I had to remind myself that the accuracy or inaccuracy of one prediction didn't necessarily reflect that of any other. Until a clear pattern was empirically established, each prediction had to be considered independently and evaluated in terms of the tectonic features and geological history of the area in question. And granting that there were errors in what the psychics said, these might have been compounded by errors in my interrogations and in my interpretation of what they had to say, or have written, and my "consensual" presentation of their view. Furthermore, it should be emphasized that the psychics did not believe that all their predictions inevitably had to come about. They felt that some of them could be averted or substantially eased in severity if enough people changed in attitude.

As I reviewed the similarities between the psychic group's predictions and Cayce's, I also noticed differences between them. Each individual had a unique quality which emphasized the varying ways by which these men and women obtained this information and which pointed to differing sources for their information. Abrahamsen, for example, said that in his meditative state he actually saw the events happen. He saw scenes of the results and at times saw the results tabulated on a large blackboard or calendar. When he told me this, I immediately wondered if he could see "behind the scenes." Could he see the actual geological processes taking place that would cause these changes? He said he sometimes did, and this led to a series of questions about geological factors that will be taken up in a later chapter. I also got surprisingly technical and scientifically interesting answers when I went through the same process with Karish.

As a result, I can state without hesitation that the spectrum of readings is geologically sound. There are internal consistencies such as the prediction of earthquake activity for certain cities and coastal areas (at definite times) and the same types of evidence support all the predictions—evidence from tectonic maps, the geological record, and recent activity. The major difference between the psychics' predictions and geological theory is the rate at which geological change takes place. The predictions call for sudden, catastrophic change while traditional geological theories call for slow and gradual change—certainly nothing like the Decade of Cataclysm. But if we recall the various sudden geological changes occasionally resulting from earthquakes that I've cited here, the psychics' predictions should cause us to scrutinize them carefully and watch for them. In the next chapter, I will give evidence that some of the psychics' knowledge of the geological processes behind tectonic maps is greater than that of the people who make such maps. I will show how they, in their altered states of consciousness, were aware of geological facts that wide-awake researchers are just beginning to discover.

CHAPTER 4

The Geological View: Catastrophism Versus Uniformitarianism

"As the ground opened up before his feet, man, in discovering that the geological doctrine of uniformitarianism was false, was also going to wonder if cataclysms had ever occurred before."

William Irwin Thompson,
At the Edge of History

"There were global catastrophes in prehuman times, in prehistoric times, and in historical times. We are descendants of survivors. We read here a few pages from the log book of the earth, a rock rolling in space, circling with its attendant lifeless satellite around a fire-breathing star moving with this its primary and other revolving planets through the galaxy of the Milky Way of hundreds of millions of burning stars, and together with this entire host, through the void of the universe."

Immanuel Velikovsky,
Earth in Upheaval

Fearsome as it may seem, there is evidence that the earth periodically undergoes cataclysmic geological changes, and that these violent events have been the major sculptors of the earth's surface. Quite independent of the psychic view, some geologists believe that the earth is about to enter another such cataclysmic period. They interpret the recently increased volcanic and earthquake activity around the world as telltale signs. The earth's internal burners seem to be heating up. Even recent geological *inactivity* in traditionally active tectonic zones such as along the African crustal plate or the central United States is viewed suspiciously. These areas could be storing energy for a really big movement.

Were the lofty Himalayas, a thousand-mile-long wall across northern India, the highest mountains in the world, created by slow and steady geological processes? Was towering Mount Everest (29,028 feet) gradually pushed up just a few inches every century? Scientists of the nineteenth century were dismayed to find that the highest rocks of these massive peaks yielded skeletons of marine animals, ocean fishes, and shells and mollusks. How are we to explain the sea floor of remote times becoming the lofty highlands of today? Instead of slowly creeping up, were the Himalayas really born in a few explosive moments? Were the Himalayas born when the Indian subcontinent *slammed*, not pushed, into the Asian mainland? Were colossal upheavals and uplifts triggered? Did it take millions of years or just a few centuries to create the Himalayas? Two opposing geological schools of thought confront each other here: uniformitarianism versus catastrophism.

Most geologists today subscribe to the only view "officially" accepted, the theory of uniformitarianism. Dr. William Stokes, in his standard text *Essentials of Earth History*, defines uniformitarianism as "the belief or principle that the past history of the earth . . . is best interpreted in terms of what is known about the present. Uniformitarianism would explain the past by appealing to known laws and principles acting in a *gradual*, uniform way through the past ages." In other words, geological features such as mountains and glaciers are supposed to have formed in the past by the same natural processes we see operating today. Ancient offshore deposits, for example, were built up in the same way that we today see water carrying loads of mud and silt to the sea.

Unlike the wide support for unformitarianism, only a few geologists, such as Dr. A. T. Wilson of Victoria University, in New Zealand, Dr. D. V. Agers of University College of Swansea, Great Britain, Dr. Peter

Vogt of the U.S. Naval Oceanographic Office, and Dr. Bruce Heezen of Columbia University's Lamont–Doherty Geological Laboratory, subscribe to the theory of catastrophism to explain some past geological events. Nevertheless, in an article appearing in the *New York Times* on January 22, 1978, geologist Stephan Jay Gould of Harvard University said, "Geologists no longer reject catastrophic change out of hand. The great channeled scablands of eastern Washington were scoured in a matter of days by gargantuan torrents issuing from an enormous glacial lake that emptied when its ice dam broke at the end of the last ice age." Stokes defines catastrophism as "the belief that the past history of the earth . . . has been interrupted or greatly influenced by natural catastrophes occurring on a worldwide or very extensive scale." The two views obviously differ with regard to the time required for change (slow and gradual versus near-overnight), and in the main component of change (constant daily small-scale events versus rare large-scale events).

In the eighteenth and nineteenth centuries, when geology was in its infancy, each school of thought had its distinguished and vociferous adherents. The uniformitarian view was proposed by James Hutton of Edinburgh in 1785 and popularized by the English geologist Charles Lyell in 1830. On the other side of the question, as reported by Stokes, the doctrine of catastrophism was supported by "many of the best minds of the nineteenth century." One of the chief nineteenth-century supporters of catastrophism was the Frenchman Baron Georges Cuvier. Cuvier, a versatile genius, backed his views with the extensive studies he made in the Paris Basin. These systematic investigations earned him the title "Father of Paleontology." (Paleontology is the study of the plant and animal life of the past based on the fossil remains found in the earth.) From his field observations, Cuvier felt that, in Stokes's translation,

> living things without number were swept out of existence by catastrophes. Those inhabiting the dry

> lands were engulfed by deluges; others whose home was in the waters perished when the sea bottom suddenly became dry land; whole races were extinguished leaving mere traces of their existence, which are now difficult of recognition, even by the naturalist. The evidence of those great and terrible events are everywhere to be clearly seen by anyone who knows how to read the record of the rocks.

Stokes points out:

> It was argued by the catastrophic school that since no man has seen a mountain come into existence or the ocean rush far inland or the glaciers form in the northlands, these events must have come about in past time in a manner unlike anything currently operating. Even though it was admitted that slow changes were taking place (such as stream erosion of banks) they appeared to be entirely insufficient to accomplish the observed results.

The catastrophists believed that the same series of catastrophes that annihilated living things was also responsible for most of the geological changes on the earth's surface. Geology at this time was trying to get on its scientific feet and justify itself. Geologists wanted to release people from the grip of Bible literalists, who taught that the earth was created in six days.

Geologists tried at all costs to avoid the rare and mysterious. For example, Charles Lyell entreated his colleagues "to exclude the miraculous from geologic explanation." Thus uniformitarianism was pursued instead of catastrophism—which brought uncomfortably to mind not only miraculous volcanic explosions but also Noah's Flood.

Thus the catastrophic school quietly faded away and their advocates devoted themselves to paleontology and glacial geology. From its free-thinking beginnings, geology then went ultraconservative. The sudden surface changes produced by catastrophes such as earthquakes and volcanoes were considered too small in size and too rare as a foundation on which to build a science. Yet to this day geology still has not explained such basic processes as mountain building, the growth of ice caps, or the widespread mass annihilation of living things.

Ironically, one early concept has become essential once again to geologists, and it might even bring the catastrophism school back with it. That concept is continental drift. In 1620 Francis Bacon first commented on the similarity between the outlines of the coasts on each side of the Atlantic Ocean, the bulge of Brazil being the reciprocal of the west coast of Africa. This observation was central to the concept of continental drift that Alfred Wegener championed in the 1930s. But until a few years ago few geologists supported Wegener's idea that the continents originally were joined together in one supercontinent, which subsequently broke up into seven pieces (the seven continents) and drifted apart to the positions they hold today. However, after a careful matching of coastlines, geological beds, fossils, and other clues, geologists realized their error and have now fully embraced continental drift.

This is quite an about-face. What was once immovable is now movable! Geologists now tell us that the continents drift at the slow and steady rate of several inches per year. Instead of horizontal or vertical internal pressure alone forcing mountains up, geologists now use continental drift to explain the origins of mountains. In the case of the Himalayas we are now told that the Indian subcontinent drifted into the Asian mainland and forced the mountain ranges up. Relying on the principle of uniformitarianism, geologists unquestioningly interpret this collision to have been a gradual one, since no slam is now being observed. But a catastrophic interpretation of this collision would work just as well. It could be argued that the original collision was violent and sudden. For example, the slowly drifting continents could have locked together, increasing potential energy, and then in a very short time, geologically speaking, quickly pitched forward to throw the mountains up. Or if pole shifts really occur, perhaps such a flip temporarily accelerated the normally slowly drifting continents, or perhaps sudden upwellings from the earth's interior temporarily accelerated them. If the collision were large enough, a chain reaction might have been created where violent movement in one direction set up and triggered violent countermovement in another direction, and so on, like a spring recoiling.

Cataclysmic interpretations may sound fantastic, but woe to those who place their trust in proclamations of the geological establishment. Ten years ago continental drift was so much nonsense! If this turnabout is not enough, in June, 1977, Drs. Amos Nur of Stanford University and Zvi Ben Avraham of the Weizmann Institute in Israel reported to the annual meeting of the American Geophysical Association that the present seven continents may not have been all that drifted over the earth's core. These scientists believe they have discovered evidence for the existence of an eighth continent, which they have named Pacifica. According to *Science News* (June 18, 1977), Nur and Ben Avraham said that the eighth continent broke up into various pieces which "collided with and rumpled the western coastline of North America, creating among other things the Sierra and Rocky Mountain ranges. Other fragments in the meantime were bumping into South America, Alaska, Kamchatka, Japan, and east Asia, also creating wide mountain ranges there. [Note how geologists use catastrophic words to describe events they feel are governed by the principles of uniformitarianism.] Besides the mountain ranges, Pacifica might have left behind little remnants of itself, which partly sank for unknown reasons into the Pacific Ocean." (This report of continental masses mysteriously sinking in the Pacific will, I am sure, bring joy to those who believe in the mythical lost continent of Lemuria, said to have been in the Pacific. Perhaps the lands which the psychics predicted will rise in the Pacific are a return of the broken remnants of this sunken continental plate.) While I can see how a collision with the eighth plate would force California's Sierra Mountains up near the coast, I wonder how a "gradual" collision would force up mountains as far inland as the Rockies. Something much more catastrophic seems required.

Again, we must ask how long it took to create the Himalayas—millions of years or just a catastrophic few centuries? Theorist Dr. Immanuel Velikovsky has long been a thorn in the geological establishment's side. In *Earth in Upheaval*, which talks about the catastrophic evolution of the earth, Velikovsky argues that the Himalayas were formed cataclysmically because they are very young in age. He says that the Himalayas couldn't have taken tens of millions of years to form because they are in fact less than one million years old.

Geologists at least agree that the Himalayas were formed *relatively rapidly* and relatively recently. For example, Stokes notes in his textbook that "the geologically rapid rise of the Alpine-Himalayan Chain is one of the most spectacular of past events." But Velikovsky cites evidence for an *extremely rapid* rise within the most recent geological period, the great Ice Age, when man's direct ancestors had already begun to move about. (It should be noted that the Ice Age was once held to have begun 3.5 million years ago, but recently geologists have redated it to have begun just 1 million years ago.) In other words, Velikovsky says people bore witness to the Himalayas rising. He notes that Swiss geologist Arnold Heim in his book *The Thorn of the Gods—An Account of the First Swiss Expedition to the Himalayas* (1939) reports finding deposits containing paleolithic (Ice Age) fossils, which would make it plausible that the mountain passes in the Himalayas may have risen three thousand feet or more in the age of man, "however fantastic changes so extensive may seem to a modern geologist." Velikovsky cites a report entitled *Studies of the Ice Age in India and Associated Human Cultures* by Dr. H. de Terra of the Carnegie-Mellon Institute and Dr. T.T. Paterson of Harvard as one long argument that the Himalayas were rising during the Ice Age and didn't reach their present heights until historical times. Tilting of terraces and lacustrine (lake) beds are given as evidence indicating a "continued uplift of the entire Himalayan tract" during the *last* phase of the Ice Age. When Arnold Heim, the Swiss geologist, also explored the mountain ranges of western China, adjacent to Tibet and east of the Himalayas, he came to the conclusion that they had been elevated even more recently. In fact, he believed they had been elevated *since* the glacial age. This means they would have risen in less than ten thousand years.

In a way, catastrophism is not inconsistent with uniformitarianism. In the uniformitarian view "the present is the key to the past," and we know that the geological changes we see in the present have not always been so slow and gradual or rare. Since the turn of this century when science began to monitor and record geological changes, we have learned of a number of catastrophic events. In 1899, for example, after an earthquake near Disenchantment Bay, Alaska, the beaches stood forty-seven feet above the sea, and a wide expanse of sea floor became dry land. Another example, already noted, is the 1964 Anchorage, Alaska, quake that tilted land either up or down in a number of areas by as much as fifty feet. The 1923 earthquake that destroyed one-third of Tokyo and most of Yokohama, killing one hundred forty-three thousand people, caused the sea floor in the adjoining Sagami Bay to sink 1,310 feet. In the 1960 Agadir, Morocco, quake, the sea floor rose as much as three thousand feet in some places. The 1939 Chile earthquake razed an area of more than fifty thousand square miles. And some quakes have produced horizontal displacements that have run for tens of miles. One can only wonder what displacements have accompanied very large quakes in sparsely populated areas (which have gone unstudied), such as in the New Hebrides Islands and Antarctica. What major movements occurred in the historical past before man began to carefully record such information? For example, what was the *maximum* movement in the colossal 1811 New Madrid, Missouri, quake, where one of the few observers noted thirty-foot fissures opening?

What are we to make of the statement by Robert O.

Castle, of the USGS, as quoted in *Science News* of January 7, 1978: "About mid-1961 . . . an area including Palmdale, Barstow, and Mojave [California] rose *abruptly* [Castle's emphasis] as much as 25 centimeters (approximately 10 inches), then gradually increased . . . during the following decade"? He went on to say that in a short time much of the uplift just as suddenly subsided. What if a ten-foot rise and subsequent subsidence occurred?

Beyond the occurrence of single great quakes and their aftershocks, is it possible that in the past an area has been repeatedly hit by large quakes? For example, Dr. J. Gilluly in *The Principles of Geology* notes that the Pleasant Valley earthquake, which rocked the almost uninhabited desert of central Nevada in 1915, produced a low cliff almost sixteen feet high where none existed before—a cliff that ran for seventeen miles along the western base of the Sonoma range. But he adds that geological studies show the total vertical displacement of this fault, which separates the Sonoma range block from the Pleasant Valley block, is much greater—at least two thousand feet. He concludes that the 1915 displacement was only the most recent activity in a series that has pushed the Sonoma Mountains above the adjoining valley. Depositional evidence from the valley indicates that the entire two-thousand-foot displacement has taken place in a very short amount of time, geologically speaking. What could quakes of 12.0 magnitude bring, especially if they hit an area repeatedly? Could such "superquake" activity explain the origin of some of the enormous faults geologists have observed? Has modern man just been fortunate in not having had to live through such violent events yet?

If a superquake of 12.0 on the Richter scale were to strike, it is not hard to imagine what it would be like. It would probably feel as if the entire earth had cracked from surface to core. A fault could open large enough to rival the Grand Canyon. The ground waves set in motion would be so high that they could blot out the skyline. A person one moment might find himself deep within a valley only to be buoyed up to a hilltop in the next moment. The vibrations from the quake center would probably be strong enough to turn telephone poles over two thousand miles away into masses of splinters. Huge tidal waves would be set off in which thousand-foot walls of water could wash over whole cities, possibly over whole states. If the water from these tidal waves were to pour into erupting volcanoes, the resultant explosions would make people think a nuclear holocaust had begun.

Surtsey is a volcanic island that suddenly came into existence in the North Atlantic off Iceland in 1963 when the earth's solid skin cracked apart and hot magma flowed up from the deep interior. This event doesn't give one a picture of slow and gradual change. In a spot where nothing but ocean had existed, a new name suddenly had to be put on the world map. One wonders how long it took the immense volcanic flows of Washington and Oregon, the most extensive of all volcanic deposits, to accumulate to thicknesses of thousands of feet and become a permanent addition to the North American continent.

In England, Dr. Hubert Lamb of the British Meteorological Office has found that an average of five eruptions as large as Krakatoa occurred in each of the past five centuries. The *National Geographic* (November, 1976) reports that Finnish geologist Vaino Auer of the University of Helsinki "has found volcanic ash layers in Patagonian peat bogs that indicate extraordinary surges of volcanic activity . . . between 0 and 500 B.C., and 7000 and 9000 B.C." More violent eruptions of "countless times" Krakatoa's power are indicated by Yellowstone National Park's miles-wide (now obliterated) caldera, by Crater Lake in Oregon, by West Germany's Eifel area, by Africa's volcanic fields, and by all of Iceland. Nor does one get a picture of slow gradual change from the ". . . *violent* events that have occurred in very recent times . . . and are still in progress," which Dr. Hanoun Tazieff told the *National Geographic* are responsible for submerging the Afar Triangle in the Red Sea area within the last six thousand years. Some of the various catastrophic activities now occurring in this rift area were discussed in the previous chapter—activities similar to those also taking place in the Gulf of California area, with its hundreds of earth tremors per day.

While we might be aware of most of the different types of geological activities that have occurred in the past, statistically speaking our sample for making judgments about the magnitude and frequency of such events in the past is still very limited. "Establishment" views and political implications aside, the geological evidence for catastrophism seems every bit as good as that offered up for uniformitarianism.

To shed light on this problem, we must take a closer look at the record from the past. The geological "book of the earth," in addition to recording earthquakes, volcanic eruptions, and mountain building, also preserves evidence of past climate and the lives of people and animals. Examination of climatic, paleontological, and archaeological evidence clearly tips the scales in favor of catastrophism over uniformitarianism to explain how the earth has come to look the way it now does. Climatic, paleontological, and archaeological evidence indicates that the earth periodically undergoes cataclysmic changes. Dr. D. V. Agers, a geology professor at the University College of Swansea in Great Britain, feels that while uniformity may persist over long periods of time, the geological record consists mainly of the effects of brief cataclysms. In his 1973 book *The Nature of the Stratigraphic Record,* Agers says, "It seems to me, from a number of recent papers (and from common sense), that the rare event is becoming more recognized as an important agent of recent sedimentation. Traditional concepts such as gentle continuous sedimentation are not adequate to explain what we see."

Immanuel Velikovsky has long championed the cataclysmic view. In *Continents in Motion*, Walter Sullivan, science editor for the *New York Times*, acknowledges modern geology's debt to Velikovsky for forcing a reexamination of old assumptions about the planet's formation. In 1950, many scientists actually threatened to boycott Velikovsky's first publisher if it brought out *Worlds in Collision*, his initial challenge to the uniformitarian view of earth's evolution. Two other strong attacks on establishment thinking by Velikovsky ensued—*Ages in Chaos and Earth in Upheaval*—which further presented his case for the cataclysmic life of our planet. While the scientific community tried to ignore Velikovsky's books, the public bought them by the millions. Over the years Velikovsky has gained a measure of respectability because his predictions, based on his theories, have proven correct. But by no means has geology as a whole greeted Velikovsky with open arms and minds. Dr. H. B. Klosterman, in an address at the 1976 Rio de Janeiro geology conference, said, "When proposed by geologists of non-catastrophic persuasion, such hypotheses are taken seriously; but when similar ideas are forwarded by less conditioned outsiders, they are regarded as evidence of lunacy simply because they violate uniformitarian dogma."

Velikovsky, a physician, has no training in geology and his work shows a certain naiveté about where the various pieces of evidence fit. Oftentimes he has included dubious or even totally incorrect scientific evidence, resulting in the bad data tainting the good. But even though he is out of his professional field, Velikovsky is a rigorous scholar and logical thinker in his effort to present a comprehensive picture of the earth's past. And since he is not schooled in geology, he has fortunately not fallen prey to geology's dogmas and myopias. Since he often doesn't know better, Velikovsky dwells on certain anomalous facts that geologists have in the past quickly swept under the carpet. Like him or not, since the publication of his first book in 1950 Velikovsky must be credited with correctly anticipating a number of scientific discoveries about our planet and the solar system, such as the emission by Jupiter of radio signals and Venus' retrograde motion.

What probably antagonized geologists most was Velikovsky holding a few dead mammoths under their noses, figuratively speaking. Velikovsky points to the perfectly preserved ancient mammoths (ancestors of the elephants) that have been unearthed in a standing position from the frozen ground in various parts of Siberia. He sees them as evidence that the earth underwent catastrophic change. These mammoths seem to have been "flash frozen" because undigested bits of temperate flora such as buttercups were found between their teeth and in their stomachs. Velikovsky cites report after report referring to the sudden extinction of the mammoths.

Dr. Frank Hibben, an eminent archaeologist from the University of New Mexico, goes Velikovsky one better and holds entire herds of mammoths before geology's face. In his book *The Lost Americans* Hibben writes:

> This was a vague geological period which fizzled to an uncertain end. This death was catastrophic and all-inclusive. What caused the death of 40,000 animals?
>
> The "corpus delicti" in this mystery may be found almost anywhere. Their bones lie bleaching in the sands of Florida and in the gravels of New Jersey. They weather out on the dry terraces of Texas and protrude from the sticky ooze of the tar pits off Wilshire Boulevard in Los Angeles. The bodies of the victims are everywhere. We find literally thousands together... young and old, foal with dam, calf with cow... The muck pits of Alaska are filled with evidence of universal death... a picture of quick extinction. And argument as to the cause must apply to North America, Siberia and Europe as well. Mammoth and bison were torn and twisted as though by a cosmic hand in godly rage. In many places the Alaskan mud blanket is packed with animal bones and debris in trainload lots... mammoth, mastodon... bison, horses, wolves, bears, and lions... A faunal population... in the middle of some catastrophe... was suddenly frozen in a grim charade.

In 1894 Dr. J. D. Dana, the leading American geologist of his day, wrote in his *Manual of Geology*: "The encasing in ice of huge elephants and the perfect preservation of the flesh shows that the cold finally became *suddenly* extreme, as of a single winter's night, and knew no relenting afterward [emphasis added]." Even Charles Darwin, who denied the occurrence of continental catastrophes in the past, admitted that the extinction of the mammoths in Siberia was for him an insoluble problem. In the classic journal of his travels (*Journal of Researches into the Natural History and Geology of the Countries Visited During the Voyage of H.M.S. Beagle Round the World*) Darwin wrote,

> What then has exterminated so many species and whole genera? The mind at first is irresistibly hurried into the belief of some great catastrophe; but thus to destroy animals both large and small, in Southern Patagonia, in Brazil, on the Cordillera of Peru, in North America up to Bering's Straits, *we must shake the entire framework of the globe.* No lesser physical event could have brought about this wholesale destruction not only in the Americas but in the entire world. . . . Certainly no fact in the long history of the world is so startling as the wide and repeated extermination of its inhabitants [emphasis added].

Velikovsky feels that the sudden extinction of the mammoths around the world and other extinctions point to the great catastrophes that befell the earth, catastrophes in which the earth actually tumbled in space, resulting in overnight climatic changes and accompanied by incredibly destructive geological changes. In *Earth in Upheaval* Velikovsky cites evidence for past catastrophic climatic and geologic changes gathered by geologists, paleontologists, palynologists, oceanographers, climatologists, and other scientists. In addition to the matter of the mammoths Velikovsky takes up:

1. The erratic boulders found around the world.
2. Instances where sea and land have changed places.
3. Caves in England jammed full of animal bones of

species that normally don't associate with each other.

4. Aquatic graveyards.
5. Hippopotamus extinctions.
6. Unusual geological deposits on the Russian plains.
7. The Ice Age in the tropics.
8. The ice-free deposits of Greenland.
9. The corals of the polar region.
10. Whale remains in the mountains.
11. Unusual tidal-wave deposits in rock fissures.
12. Submerged forest beds in Norfolk, England.
13. Mountain thrusts in the Alps.
14. Anomalies associated with the formation of the Himalayas.
15. The unusual deposits of the Sewalik Hills of India.
16. The geology of Tiahuanaco in the Andes.
17. The formation of the Great African Rift.
18. The origins of the Sahara.
19. The parallel craters of the Carolina bays.
20. Geological activity along the Mid-Atlantic Ocean Ridge.
21. The cause of ice ages.
22. The earth's shifting poles.
23. Continental drift.
24. The earth's changing orbit.
25. The earth's rotating crust.
26. The earth's shifting axis.
27. The rapid evaporation of oceans.
28. Rapid condensation and the formation of oceans.
29. Magnetic-pole reversals.
30. The anomalous occurrence of volcanoes, earthquakes, and comets.
31. The formation of glacial Lake Agassiz.
32. The start of Niagara Falls.
33. Changes in the Rhone Glacier.
34. Changes in the flow of the Mississippi River.
35. The lakes of the Great Basin as related to the end of the Ice Age.
36. Tree rings and dating of past events.
37. Dropped ocean levels around the world.
38. The origins of the North Sea.

39. The distribution of coal deposits.
40. Fossil deposits in Cumberland Cavern, Maryland, in northern China, in the tar pits of La Brea, California, and in Agate Spring Quarry in Florida.

Velikovsky's interpretation of this evidence sometimes is farfetched and inappropriate, and he often disregards well-established geological datings, but many of Velikovsky's points are nevertheless well taken. The geological establishment still cannot adequately explain many of these topics, especially mountain building, biological extinction, magnetic-pole shifts, and glaciation in areas close to the equator such as in Brazil and India.

Velikovsky also cites evidence from the archaeological record that indicates that natural catastrophes were responsible for the downfall of certain ancient Eastern civilizations. He says the greatest of these catastrophes produced seismic disturbances of such unusual severity and extent that it caused the downfall of the Middle Kingdom in Egypt. He tells in *Earth in Upheaval* how

> cities were overturned; epidemics left dead piled in common graves; the pursuit of arts and commerce came to an abrupt end; empires ceased to exist, strata of earth, dust, and ashes yards thick covered the ruined cities. In many places the population was annihilated, in others it was decimated; settled living was replaced by nomadic existence. Climate changed.

Velikovsky synchronizes the closing hours of the Middle Kingdom with the biblical Exodus, where sea, land, and sky were in an uproar. He bases his arguments for natural catastrophes befalling many peoples around the world on archaeological site reports, historical texts, classical literature, epics of northern nations, sacred books of the Orient and the West, and tradition and folklore of primitive peoples. For example, he relates how Sir Arthur Evans, the famed excavator of Crete, found evidence for widespread catastrophic destruction in the

Mediterranean. In his archaeological report dealing with his excavation of the palace of Minos, Evans speaks of a "great destruction" that befell Knossos and the Minoan kingdom. He writes that the phenomenon "conclusively points to a seismic cause for the great overthrow that befell the Palace and surrounding town. . . ." Dr. Spiro Martinatos feels that "the catastrophe of Late Minoan I was fatal and general throughout the whole of Crete." Martinatos believes Crete received an "irreparable blow" as a result of natural causes. "A normal earthquake, however, is wholly insufficient to explain so great a disaster," he writes. Velikovsky also quotes a coast guard officer who reported that "some tremendous catastrophe had raised a section of the island far above the level which it occupied when the city of Knossos existed." Velikovsky argues that at the same time Crete was shattered by a violent catastrophe that destroyed the cultural and political system of the Minoan kingdom, Egypt's Middle Kingdom (only 400 miles from Crete) was terminated (approximately 1800 B.C.) and about this time Troy III's fortification wall fell, and the island of Thera (Santorini) in the Aegean (only 100 miles from Crete) exploded with almost unimaginable fury.

Velikovsky's argument was made in the 1950s and new research is strengthening it. In the early 1970s archaeologists excavating at the island of Thera off the Greek coast came to realize how great was the civilization that flowered there, the key role it played in Aegean politics, and how it was totally destroyed by a volcanic eruption that geophysicists consider to have been the largest in human history. The volcano's entire crater was blown off. Whether this colossal eruption was related to the seismic disturbances that occurred in other countries about this time is not yet definitely known. But Velikovsky has laid out a potentially valid hypothesis.

Digressing for a moment and taking Velikovsky's free-wheeling thinking a step farther, it is interesting to note that about the time he feels destruction was rained on the civilizations of the Mediterranean and the Near East (1800 B.C.), the first civilizations of America sprang up. At approximately 1800 B.C., the Chavin civilization of Peru and the Olmec civilization of Mexico began. In the last few years, the most eminent archaeologists in the world have suggested the Near East as the origin for the American civilizations. For example, Dr. David Kelley of the University of Calgary in *The Alphabet and the Ancient Calendar Signs* points to the many similarities between the Near Eastern and the Mayan calendric systems. And in the March, 1975, issue of the *American Anthropologist* there is an article entitled "The Transpacific Origin of the Meso-American Civilization." Dr. Betty Meggers of the Smithsonian Institution lists extensive evidence for the Olmecs in Mexico having found their origins in the Near and Far East. Meggers even calls attention to the similarity between the Minoan Linear A script of Crete (dated between 1700 and 1600 B.C.) and those of Olmec rock carvings. Note also the many myths about these ancient Indian groups having been visited and helped by fair-skinned people who came in boats from the East. One is led to wonder just how big a role catastrophes have played in influencing the course of humanity.

Velikovsky also makes many references to cataclysms discussed in the literature of antiquity. He goes far beyond the theme of the great flood, which is common to cultures around the world. Velikovsky points out that Heraclitus (540–475 B.C.) taught that the world is destroyed by conflagration every 10,800 years. Other Greek scholars taught that the earth periodically undergoes destruction by combustion and deluge. The sacred Hindu *Bhagavata Purana* tells of four ages and of *pralayas*, or cataclysms, in which various epochs humanity was nearly destroyed. References to ages and catastrophes are found in the *Zend Avesta*, the sacred scriptures of Mazdaism, the ancient religion of the Persians. The ancient Chinese encyclopedia *Sing-li-ta-tsiuen-chou* discusses the general convulsion of nature, and one passage relates that "in a general convulsion of nature the sea is carried out of its bed, mountains spring out of the ground, rivers change their course, human beings and everything are ruined, and the ancient races effaced."

The tradition of world ages that ended in cosmic cataclysm is also found in the Americas among the Incas, Aztecs, and Mayans. The Mayan calendric disk makes specific reference to these catastrophes. These are noted as times of flood, and times when the earth breaks in many places and mountains fall. The early Spanish explorers in the Americas repeatedly recorded such stories as given to them by Indian scholars. Velikovsky in *Worlds in Collision* notes the *Manuscript Troano* of the Mayans, which describes a catastrophe during which

> the ocean fell on the continent, a terrible hurricane swept the earth, volcanoes exploded, tides swept over mountains, and impetuous winds threatened to annihilate humankind, and actually did annihilate many species of animals. The face of the earth changed, mountains collapsed, other mountains grew and rose over the onrushing cataract of water driven from oceanic spaces, numberless rivers lost their beds, and a wild tornado moved through the debris descending from the sky. The end of the world age was caused by Hurakan, the physical agent that brought darkness and swept away houses and trees and even rocks and mounds of earth. From this name is derived "hurricane," the word we use for a strong wind.

After reading the vast literature Velikovsky cites, one can only conclude that the authors could not all have been caught up in exaggeration along the very same themes, but rather that at least some of the literature accurately records the fact that humans have watched the earth undergo periods of cataclysmic change in the recent past. If the 220-foot tidal wave that appeared off Alaska after the 1964 quake struck the coast, would it not give a person the impression that the "ocean is falling on the land"? Recall the quotations from eyewitnesses to earthquakes or tidal waves with which this book began. Consider also

Walter Sullivan's vivid description in the *New York Times Magazine* of the powerful New Madrid, Missouri, earthquake that hit along the Mississippi River on December 15, 1811.

> Shortly before 2 A.M., there was a creaking and groaning of the timbers in houses and cabins. Then a fearful crashing as furniture was overturned, chimneys crumbled and china fell to the floor. Fleeing from their homes, the people shivered outdoors through the rest of the night.
>
> At dawn came a low rumbling, then catastrophic upheavals of the earth. The landscape rippled in waves, like a ground swell at sea. The tops of trees intertwined, then were wrenched apart as the earth convulsed beneath them. In some places, great cracks appeared in the ground, then closed. The Mississippi, thrown from its bed onto the hillsides, swept away entire forests. Boats were overwhelmed or thrown far inland. Downstream, in Tennessee, the river was rerouted, creating Reelfront Lake, 10 miles long.

Compare this description with the ancient literature. Those accounts not only describe large-scale earthquakes but volcanic eruptions, tidal waves, hurricanes, and mountains rising or falling—the type of cataclysmic period that modern man has been fortunate to have missed so far.

Unfortunately geologists, geophysicists, and seismologists do not seem to be interested in studying evidence from paleontology, climatology, archaeology, ancient literature, and folklore to get a better picture of the potential energy the earth can release at one time. Nor are they willing to read Velikovsky's books with open minds in search of clues for guiding their own research. The reader who has come this far has already taken a bigger step than 99 percent of our earth scientists. It won't do for our geologists one day to apologize to us and say they were wrong concerning cataclysmic geological change.

Why won't it do? There just may not be much of humanity left to hear the apology.

Yet there seems to be hope. Some earth scientists are loosening their rigid mind-set somewhat. They now admit that all change is not agonizingly slow. Continental drift and sea-floor spreading having been accepted, geological change is now viewed as occurring much faster. The crust of the earth is no longer viewed as a rigid shell subject only to extremely gradual and piecemeal change from weathering cracks and gentle bulges. Instead, entire sections of the earth called crustal plates (themselves rigid—both continental and ocean plates) are free to move about, with the margins of some plates continually being consumed in deep ocean trenches while new crustal material continually wells up from rifts in the ocean floor. The Atlantic sea floor is said to be spreading and the Pacific sea floor shrinking, while the continents drift apart with occasional collisions, and with one continent possibly having been lost in the process.

Even though the new theories greatly reduced the time needed to effect change, geologists still cling to their concept of slow, uniform change—and it must be said that recent studies have shown that the current rates of these movements are presently relatively slow and steady. But with the new mobile crust the potential for rapid change is there and the questions now are: how fast can geological change take place, and how fast have changes taken place in the past? Have all plate movements been as slow and steady as those we see today or have some movements taken place quite rapidly? Is plate tectonics—at whatever rate—in fact responsible for mountain building and for the opening of the ocean basins (sea-floor spreading), or are geologists still missing a big piece of the puzzle?

The organizing committee of the American Geodynamic Project, part of the International Council of Scientific Unions, notes in its official publication that all mountains do not occur as long ridges paralleling present or former plate boundaries, as the Himalayas and the Andes do. The Rocky Mountains of the western United

States and the Brooks Range in Alaska are not located along plate boundaries and are thus not explainable by plate theory. The highest peak in North America, Mount McKinley in the Alaska Range, lies 250 miles from the edge of the American Plate. The committee also observes that plate theory cannot explain inland volcanic activity such as that which produced the San Juan Mountains of Colorado. They ask why there are swarms of seamounts on certain parts of the Pacific floor instead of the long, southeast-northwest lines of islands one would expect to find if the plate had drifted over a hot spot. They ask why there are submarine ridges like the Bermuda Rise, which seems unrelated to plate boundaries. Walter Sullivan in *Continents in Motion* notes that the committee is particularly bothered by the fact that andesite, the volcanic rock found in the Andes and other mountains formed by continental drift, is virtually absent in the Alps, and that the cause of great uplifts is still a puzzle. Sullivan writes that "the new evidence for massive horizontal thrusts has diverted attention from indications that there have also been large vertical movements." Such vertical uplifts have affected continental areas like the Tibetan and Colorado plateaus and the mid-ocean ridges.

Note as well that the recent extensive photography of the mountains and volcanic landscape of Mars does not show features that support slow movement of long rigid sections of crust. Perhaps the right explanation is to be found in the long-held suspicion of some geologists that some form of upward flow in certain parts of the earth's interior is responsible for most geological change? Could plumes of hot semimolten rock *suddenly* rise to form mountains, just as they did to give birth to the volcanic island of Surtsey? Also, note that geologists still have not come up with a good explanation for what sets the continents in motion and keeps them going. Do key geological changes occur uniformly or catastrophically? You must be your own judge because geologists are really only guessing. Their guesses are based on their training and modified by current geological concepts, but whether these concepts are good or bad is hard to say. The geological about-face in favor of continental drift is about as big as the about-face in Columbus' time from a flat earth to a round one.

We must also ask ourselves what role modern man's activities play in geological change. When modern man backs up huge bodies of water in dams, drastically lowers water tables, sets off atomic blasts, dumps atomic and other lethal wastes in deep wells, pumps oil from the ground, and then drives the last barrels of oil out by pumping soap suds into the ground, is he adding to the possibility of cataclysmic geological change? Can these local events somehow cause an increase in both local and global geological activity? Are there connections in the release or blockage of geological energy between different parts of the world?

While we don't yet know the far-reaching effects of underground atomic blasts such as those detonated in Nevada or Siberia, we do know that when the U.S. army disposed of lethal waste material by dumping it down deep wells outside Denver in the mid-1960s, Denver suddenly started to undergo its first recorded earthquakes. Seismologists at the Colorado School of Mines recorded more than seven hundred and ten earthquakes. Overnight Denver's building codes had to be changed to provide a margin of safety from the new earthquake hazard. Local residents sued the army over damages from the quakes. Despite denials by the army's Rocky Mountain Arsenal of any connection between their dumping activities and the quakes, and following much study by geophysicists in 1966, the army agreed to stop its disposal program. After this the earthquakes stopped as suddenly as they began. In a similar situation, USGS geologists in an experiment at the Rangely oil field in northwest Colorado injected water through the field's wells and, by varying the pressure, the *National Geographic* reports, "they showed they could turn mini-earthquakes on and off as if by a faucet." In November, 1972, these USGS geologists forced water down four of Chevron's wells. A series of mini-quakes soon began and did not stop until March, 1973. Small quakes occurred at the rate of fifteen to twenty per week. Then the scientists pumped water out of the wells and almost immediately earthquake activity ended.

On the same theme, a rare quake occurred in Michigan after oil drilling began there. The subsidence and slumping of the ground at Los Angeles area oil fields has also triggered small earthquakes. Think of all the new oil fields being opened around the world in places never drilled before. Recently major fields in China, offshore Japan, Indonesia, Africa, Alaska, and the North Sea have been added to traditional producing areas in North and South America, Eastern Europe, and the Near East.

But oil is not the only thing man pumps from the ground. One can only imagine the far-reaching effects resulting from water being drawn up from deep underground reservoirs. Also of interest is that a higher incidence of earthquakes has been observed in areas with large dams. Recently, Desiree Stuart Alexander and Robert Mark, two USGS geologists, reported evidence that high dams back up enough water to trigger earthquakes. In a sample of nineteen dams, they noted a worldwide correlation between the depth of water behind a dam and the frequency of earthquakes in the immediate vicinity. Included in these data was the 6.1 magnitude quake of August 1, 1975, that damaged Oroville, California, which is located below the huge eastern Oroville Dam, completed in 1968, on the Feather River. There is no telling what the ultimate effects of man's increasing activity in modifying the natural environment will be. Man himself may be cocking the hammer for the earth's next cataclysmic blast. And it may not be only by these activities.

CHAPTER 5

Inside the Quake: A New Predictive Technique

> "What determines the soundness of a hypothesis is not the way it is arrived at (it may have been suggested by a dream or a hallucination) but the way it stands up when tested, i.e., when confronted with relevant observational data."
>
> Carl G. Hempel
> *Aspects of Scientific Explanation*

Psychics and some geoscientists, such as Dr. Charles Richter, deriving their information independently, point to the possibility of a catastrophic earthquake generation. There is geological credibility behind many of the psychic predictions made about areas singled out for destruction. Geological maps and the geological record tend to support many of the *specific* predictions psychics have made for different parts of the world. But there is much more to support the psychic view. I not only asked the psychics I worked with for their earth-change predictions, I also asked them to look "behind the scenes" and try to discern *the actual geological processes and mechanisms that would cause the destruction they foresaw*. I went beyond the question of whether these changes took place quickly or slowly. And I didn't care if the general cause of these events was continental drift or sea-floor spreading or something else. Instead, I was seeking to determine the local physical factors involved. I was after an exact description of what happens just before an earthquake occurs or a volcano erupts. How does the ground change? Are there any warning signs? I felt that if I could discover some of the specific physical factors behind such changes, then scientists could learn to make their own predictions.

As soon as I began to question my psychic group on this, I very quickly saw consistencies between what they had predicted and what they were describing as going on internally at the time. And there was also a clear correlation between the geological mechanisms they saw and the many theories the geologists hold to explain such changes. But caution had to be used here because geological theory is changing. Yet where discrepancies have arisen, new facts constantly seem to support the mechanisms of the psychics over those of the geologists. For example, the anonymous government geologist who analyzed the Cayce earth-change readings stated that "whereas the results of recent research sometimes modify, or even overthrow important concepts of geology, they often have the opposite effect, in relation to the psychic readings, in that they tend to render them the more probable."

This is just what I found during the time I was writing this book. Each month new professional papers based on new geological facts were published that directly supported the theories I had collected from my psychic group. By looking beyond the predictions to search for the geophysical mechanisms involved, my psychics have anticipated several new discoveries about the nature of earthquakes and volcanic eruptions.

Like geologists, the psychics saw the constant need for the earth to release the internal pressure that builds up from the molten rock within it. Geologists and psychics both talk about how earthquakes and volcanic eruptions are two of the ways the earth releases its internal pressures. Cayce spoke of "upheavals in the interior of the earth" as being the direct geological cause for the earth's surface physical changes.

But my psychics went a step further than the geologists and described how the release of pressure in one area of the world can be related to the release of pressure in another area. These psychics talked about how certain areas of the world are connected by underground passages or linkages. This is consistent with these same psychic predictions of certain cataclysmic events triggering other cataclysmic events. The key warning sequence—earthquakes first in India, then Japan, then Mount Vesuvius erupts, then Mount Pelée erupts, and finally earthquakes on the west coast of North America—is a good example. The very similar eruption records of Vesuvius and Pelée have already been cited, but recently more evidence for interconnecting underground pathways being involved in the release of the earth's internal pressure has come to light.

Physicist John S. Rinehart of the National Oceanographic and Atmospheric Administration has been comparing the activities of several geysers with earthquake records dating back over a century. Yellowstone's Old Faithful, he points out, speeds up its average time between spouts from two to four years before a major earthquake within sixty miles of itself and begins slowing down shortly after the shock. In Rinehart's view, the variation in Old Faithful's timing was connected with Alaska's Good Friday quake in 1964. Rinehart feels the stresses building up in Alaska traveled through the earth at a rate of from three to six miles per day, eventually reaching as far south as Yellowstone Park in Wyoming. There he feels they put increasing pressure on Old Faithful, and as a result the geyser began spouting with increasing frequencies until the quake finally relieved the strain and allowed Old Faithful to resume its normally slower pace.

In *Continents in Motion*, Peter Vogt of the U.S. Naval Oceanographic Office is quoted as believing that the surge of molten material from one of the earth's plumes (hot spots) "may travel through channels in the floor of the lithosphere [rigid crust]" and "that a horizontal plumbing system underlies the Mid-Ocean Ridge." Vogt has found correlations between past molten surges off Iceland and eruptions from the plume under the far distant island of Hawaii. In 1972 Vogt wrote in *Nature* about the occurrence of intensified and widespread volcanic eruptions. Vogt believes that such activity occurs periodically, with revolutionary effects on the development of life on land as well as in the sea.

Little is known about the immediate cause of earthquakes. Most geologists believe that the apparent cause is the sudden breaking of rock masses stressed beyond their strength, with sudden movement along faults. Geologists believe this because such movement is enough to cause the shaking and fault displacement that has been observed during a large number of quakes. Yet the great majority of earthquakes are not accompanied by visible fault displacement (though such displacement may occur below the surface).

In December, 1972, when I began my work with Abrahamsen, nothing definitive beyond these few factors was known about the inner working of earthquakes. Since then, trying to get a better understanding of what might be going on, geologists have intensively studied highly faulted areas, the most likely locations for earthquakes, mapping and measuring everything possible, hoping to identify when conditions were right for a quake. One approach consists of measuring the buildup of stress in the ground by measuring the deformation of rocks in bore holes, or recording rock strain via special gauges. Other scientists try to estimate stress by measuring the local fluctuation in the earth's magnetic or electrical fields, or else through changes in gas emission. Still another approach consists of measuring actual changes in ground surface deformation or tilt and/or micro-movements via precise laser beams.

Despite this intensive research, we still don't have a clear understanding of the inner working of earthquakes or how to forecast their occurrence. The most promising observation made so far is of a slight swelling of the ground that sometimes precedes earthquakes and volcanic eruptions. At this very moment the ground around Palmdale, California (the Palmdale Bulge), through which the San Andreas Fault runs, is being watched. In an early reading, long before I began my research with Abrahamsen, he talked about "swelling" as a precursor of earthquakes. During one of our work sessions I asked

Abrahamsen about "swelling" and
geologists that swelling often occurred
Then Abrahamsen went a step beyond
and told me *why* the ground swelled. He
swelled because of the formation of "voids,"
pockets within the rocks below ground. Abraha
went on to say that a lowering of the water table both
aggravates this condition and accompanies it.

Soon after I finished my earthquake readings with him in January, 1973, I was pleasantly amazed when I read an article in *Science* (August 31, 1973) by Dr. Christopher Scholz, senior research associate at the Lamont-Doherty Geological Laboratory at Columbia University, Dr. Lynn Sykes, professor of geology at Columbia University and head of the seismology group at Lamont-Doherty, and Yash Aggarwal, a graduate student at the observatory. The article concerned earthquake prediction, and it described how rock dilatancy and water diffusion explained a large class of phenomena pertaining to earthquakes. Rock dilatancy is the opening of tiny cavities or cracks in rocks subjected to great pressure. The Columbia geophysicists told how they were able to predict a number of earthquakes successfully on the basis of the "tiny cavities" and cracks that form in rocks just prior to their fracturing (rock dilatancy), a subtle change also involving water saturation. Abrahamsen's terms "voids" and "tiny pockets" matched the "tiny cavities" and "cracks" the geophysicists noted forming before quakes. Likewise, his psychic perception of changes in the water table accorded with the change in water saturation the geophysicists noted before quakes.

The Columbia geophysicists were led to their discovery by Russian scientists, who marked a temporary change in the time relationship between the different components of seismic waves—vibrations that pass through the earth—before earthquakes. The Russians learned that the longer the period of abnormal wave velocity between the wave components, the larger the eventual tremor. For example, it is estimated that this change can occur as long as ten years before a magnitude 8 quake, a year before a

...itude 7 quake, and a few months before a magnitude 6 quake. The Columbia group related this wave-velocity change to rock dilatancy. They said that long before an earthquake, cavities first open in the crustal rocks, permitting trapped water to leave the rocks. Paradoxically, the rock's strength temporarily increases as a result of the drop in fluid or water pressure. This enables the rocks to resist fracture and the quake is delayed. The velocity of one of the seismic-wave components decreases because it does not travel through the new open spaces as fast as it does through the previously solid rock. Eventually, though, ground water begins to seep into the new openings in the dilated rock, seismic-wave velocity returns to normal, and the diffusing water increases the rock's fluid pressure, which leads to a quake.

Because the formation of tiny cavities in a rock can increase its volume, dilatancy may explain the ground swelling or tilting that precedes some quakes. The Japanese, for instance, noted a two-inch rise in the ground as long as five years before the major quake that rocked Niigata in 1964. In addition to the ground sometimes swelling before a quake, the local water table falls as ground water diffuses through (fills) the new openings in the rock. In 1975 Chinese seismologists successfully predicted several earthquakes by measuring the level and color of well water. Dr. Frank Press, president of the American Geophysical Union, said in a *Los Angeles Times* interview that thousands of lives could be saved in southern California if the state would learn the earthquake prediction techniques used successfully in the People's Republic of China. Press said that two months before the 1975 Haicheng quake a bulge in the earth's crust was discovered. There also was unusual animal behavior, and a number of wells became muddy and began to bubble. In 1976 a group of Japanese amateur seismologists belonging to the Namezu No Kai (Catfish Club) predicted two moderately strong tremors in the Tokyo-Yokohama region (May 13 and June 6). They made their prediction by carefully gauging the water level in numerous observation wells. An Associated Press report quoted a club spokesman, Yasuki Oki, as saying that "the water table begins dropping several hours before the main shock" and that the club "has experienced the same pattern in 20 cases, so we believe we can apply this technique to correctly predict earthquakes—just like weathermen forecasting weather."

Not only does rock dilatancy explain crustal swelling and the drop in the water table before quakes, but scientists now believe dilatancy is at the heart of other quake-associated phenomena such as changes in the local magnetic field, changes in the rock's electrical resistance, and changes in radon (a radioactive gas) over a quake area. In his unique way, Abrahamsen had described to me the mechanism behind earthquakes, a mechanism scientists now recognize as central to most of the physical phenomena associated with earthquakes.

Dilatancy adds to the credibility of some of the earthquake predictions Abrahamsen and the other psychics made. For example, recall that the psychics said that the area between the Gulf of California and the Imperial Valley would be one of the first to be hit by destructive quakes in the United States. Yuma, Arizona, lying just fifty miles north of the Gulf of California, is at the center of this zone. Ray Elkins in particular has talked about the severe quakes Yuma will receive. So it seems significant to note an article appearing on July 3, 1977, in the *Arizona Daily Star* announcing that the lower twenty-one miles of the Colorado River (before it empties into Mexico) "just south of Yuma will soon be dry as the U.S. Bureau of Reclamation will no longer release water into the channel." The water is being diverted as a result of the Colorado River Basin Salinity Control Project. Dr. Ron Olding, a biologist with the Arizona Game and Park Department, was mainly concerned with what will happen to the area's wildlife when, he said, the "basin's water table will begin to drop." If Olding had been reading seismology journals before he made this statement, he would probably have been shocked to learn that the severe drop in the water table may pose a severe threat not only to the area's wildlife, but also to the people in the area—because it is often an earthquake precursor.

Ironically, while officials in one government agency are preparing to deplete underground water from one area, officials in another agency are pumping underground water into another area. Based on the Rangely oil field experiments, where the injected water was used to turn quakes on and off, USGS geologists B. Raleigh and James Dietrich propose to control potentially dangerous fault areas by pumping water into the ground, thereby "greasing" the unstable rock layers and permitting the built-up pressure to release itself in microquakes—that is, man-made minor earthquakes. Raleigh and Dietrich proposed to do this along the entire length of the San Andreas Fault.

Dilatancy and dropping water tables explain why most members of my psychic group said that drought often signaled the coming of a quake. The minor earthquakes that struck San Francisco in 1977 may offer a small demonstration of the vast destruction psychics foresee for the city, a demonstration explained by the drought-lowered water table. In 1976 northern California began to experience a drought that is still going on at this writing (1977). In some areas the water table is dropping at the rate of six feet per month. The drought began when winter storms missed California in 1976; 1977 may mark California's worst drought on record. Crop losses have been colossal and in the spring of 1977 the San Francisco Bay area began rationing water. The surrounding counties, to quote one official, were in a "drought crisis of unprecedented proportions . . . if it doesn't rain we'll have mud in our reservoirs by September." On January 8, 1977, after the drought was well under way, a series of eight earthquakes struck the Bay area. The quakes were centered about twenty miles east of San Francisco in the Walnut Creek area. According to the University of California Seismograph Center, one of the quakes, 5.0 on the Richter scale—meaning it was capable of causing considerable damage—was the strongest in the area in more than ten years. On June 20, 1977, while the drought was still going on, San Francisco was jolted by a 4.2 quake.

Not only did Abrahamsen envision the dilatancy mechanism—with associated swelling and water-table changes—but he also pointed to a specific variable that scientists seem to have missed: temperature. In one of the readings Abrahamsen detailed a correlation between earthquakes and subsurface temperature changes. He said that before an earthquake, 200 feet below the ground there is a gradual 15 degree F. temperature increase. After this relatively slow increase, a second, more rapid temperature increase occurs. This second increase of 15 to 20 degrees F. occurs relatively rapidly just a few days before the actual earthquake. This makes for a total temperature change of 30 to 35 degrees F. before an area is struck by a quake. Abrahamsen also warned that the longer the initial temperature rise takes, the larger the earthquake that eventually will follow. He added that a longer warning period could be obtained by taking the earth's temperature at depths greater than 200 feet. For each 100 feet in depth beyond the original 200-foot minimum depth, a few more days of warning could be obtained, up to a maximum warning of about 20 days by checking the temperature at 1,200 feet.

HYPOTHESIZED PRE-EARTHQUAKE SUBSURFACE TEMPERATURE VERSUS DEPTH RELATIONSHIP

*Subsurface Depth, feet**	*Potential Period of Advanced Warning, days*
200	1
300	2
400	4
500	6
700	10
800	12
900	14
1,000	16
1,100	18
1,200	20

* At which anomalous temperature is noted.

* * *

As far as I am aware, based on a series of phone calls to the USGS officer in California, no data have been collected on subsurface temperature changes before earthquakes. Such data could easily be collected from abandoned oil wells and mine shafts in earthquake-prone areas or from wells drilled just for this purpose.

According to the laws of psychics and some of the data the Columbia geophysicists gathered, Abrahamsen's temperature earthquake prediction factors make sense. Boyle's law says that *pressure* multiplied by *volume* is equal to a *constant* multiplied by *temperature* ($PV = RT$). Keeping in mind that the Columbia University research defined rock dilatancy as the opening of myriad cavities within rocks as a result of *increasing* pressure, and that eventually the fluid pressure within the rock increases to the point where the rocks fail and an earthquake occurs, we can say that in an earthquake we are dealing with a condition of increasing pressure. But the Columbia group also emphasized that laboratory studies show that the rocks involved undergo an *inelastic* volumetric increase prior to failure. In other words, the rock doesn't necessarily increase its outer volume and swell but rather it increases its inner volume by developing added cavities. So during a quake we get an increase in pressure while volume remains constant. Applying Boyle's law, an increase in pressure while volume remains constant would result in an increase in temperature. This is just what Abrahamsen predicted. The data also show that the longer the dilatancy period, the larger the resulting quake. This directly supports Abrahamsen's statement that the longer the initial temperature rise, the larger the resultant quake.

Thus Abrahamsen has given science a new earthquake prediction technique of great potential, one by which the time and severity of a quake can be calculated. Measuring the temperature change before quakes would be the simplest and cheapest procedure. If the temperature hypothesis stands up, we would have to credit a psychic with giving science its *first* universal prediction technique. I say first because changes in the other quake signs discovered so far may not always occur. Abrahamsen's would be a universal technique because temperature is a basic variable of all phenomena. An invaluable tool for the prediction of earthquakes may be in the offing, one that will permit orderly preparation and timely evacuation before an impending quake.

CHAPTER 6

The Ice Age Cometh

Is the world's climate trying to tell us something about earth changes? Listen to this: ". . . a new climatic pattern is now emerging. We believe this climatic change poses a threat to the people of the world. Its direction . . . indicates major crop failures almost certainly within the decade."

This was the unanimous verdict of an international group of climate experts that gathered in Bonn in May, 1974, at the behest of the International Federation of Institutes for Advanced Study. They concluded that this imminent climatic change meant colder temperatures in the north and increasing variability in the general weather, with more frequent and severe storms.

Since 1974, more and more experts have been voicing their concern about our changing global weather patterns. Meteorologists find that, worldwide, the atmosphere has been growing gradually colder for the past three decades. Since the 1940s the mean global temperature has dropped about 2.7° F. The year 1976 will go down in history as a year to remember. The South Pole had its coldest year on record, with an average daily temperature of -58° F. From April to September the daily average was -77.3° F. Siberia broke its low-temperature record. Also in the USSR, snow fell farther south than in any winter on record. For two-thirds of the United States, the 1976–77 winter was also the coldest on record. Upper New York State was almost buried under a total of twenty-six feet of snow. A single snowfall of fifteen feet brought the city of Buffalo to a standstill. A freak cold front in Florida reduced the citrus crop by 14 percent and snow even fell briefly in Miami. The cities of

Chicago, Pittsburgh, Jackson, and Nashville experienced many of their lowest recorded temperatures ever. Cincinnati averaged 20 degrees below normal, while Indianapolis recorded an all-time low of –18° F. The National Climatic Center in Asheville, North Carolina, said that January, 1977, was the coldest month since at least 1800 in most of the Plains states and the East. A preliminary study indicates that the 1976–77 winter may have been America's coldest since the very founding of the Republic, and as this book goes to press, the 1977–78 winter is also setting new records. The global temperature trend shows no indication of reversing and some experts believe that the trend is even now accelerating. In fact, some feel a new ice age is imminent.

In addition to decreasing average temperature, our cooling climate is influenced by other factors:

1. The expanding Arctic. In 1971 the area covered by ice and snow suddenly increased by 12 percent, an increase equal to the combined area of Great Britain, Italy, and France. This increase has persisted ever since, and some experts think the ice is on the move again. For example, areas of Baffin Island in the Canadian Arctic that were once totally free of snow in the summer are now covered year round.

2. The great ice mass of the Antarctic, where average ice and snow cover is always high, is also expanding. In one year, 1966–67, the ice mass grew by 10 percent.

3. The North Atlantic is "cooling down about as fast as an ocean can cool," according to the climatic experts who gathered in Bonn in 1974. For example, sea ice has returned to Iceland's coast after more than forty years' absence.

4. There are also widening caps of cool air at the poles. There has been a noticeable expansion of the great belts of high-altitude dry polar winds—the so-called circumpolar vortex that sweeps from west to east around the top and bottom of the world.

5. Glaciers in Alaska and Europe that were retreating until 1940 have begun advancing again.

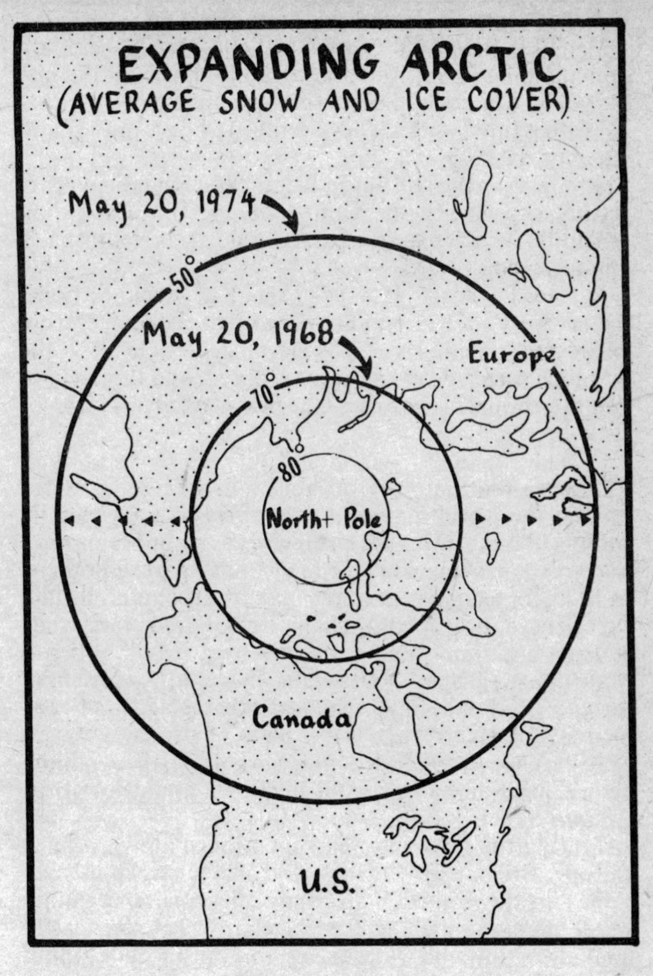

adapted from National Oceanic and Atmospheric Adm. and Time

6. A species of heat-loving snail has reportedly vanished from the forests of central Europe.

7. The winters of the Northern Hemisphere are growing longer. Satellite photography shows that in just six years, winters in the Northern Hemisphere increased by almost three weeks. They averaged 84 days in 1967 and 104 days in 1973.

8. The growing seasons in Britain and Scandinavia are getting shorter. In the United States, the heat-loving armadillo, which once ranged as far north as Nebraska, is migrating southward.

9. On February 15, 1978, the Kitt Peak National Observatory in Tucson announced that the temperature on the sun ominously fell by eleven degrees in 1977, the first time such a drop has been recorded, and they believe that this drop will increase as the next eleven-year sunspot cycle begins.

The current extreme variability in weather patterns around the world may be a side effect of the generally cooling climate. Over the last few years record rains have soaked some areas while droughts have parched others. In 1976, for example, as noted in the first chapter, Britain experienced its worst drought in five hundred years, while at the same time some parts of Africa (Mali, Senegal, Mauritania, Upper Volta, and Nigeria) experienced torrential rainfalls that spawned great plagues of rats, locusts, and caterpillars. Other parts of Africa, however, were parched by fierce drought. Since 1970, three million people have died of drought in the six countries of the African Sahel.

During the same few years mentioned above, China, Europe, Sri Lanka (Ceylon), and the Mississippi River Valley in the United States have all experienced their worst floods. Recalling the correlation between earthquakes and lowered water tables, one must wonder about the earthquake implications of the record droughts occurring from Brazil to India, from Australia to Africa and the Plains states of America (where a return to the disastrous Dust Bowl conditions of the 1930s is possible).

It is frightening to note that those parts of the United States where drought is worst at the time of this writing (summer 1977) are also some of the very areas of highest seismic risk. For example, the Pacific Northwest, from eastern Idaho down to southwestern Utah, and the South Carolina coast are zones designated as likely to receive major damage, according to the *Proceedings of the Fourth World Conference on Earthquake Engineering*, a conference held in Santiago, Chile, in 1969.

According to *Time* (June 24, 1974), many scientists believe that the bizarre and unpredictable weather patterns of the past several years are part of a "global climatic upheaval" leading to another ice age. It is believed that lower temperatures could produce a generally wetter and less stable climate, one marked by storms, floods, and freezes. For example, the widening of the polar cap and expansion of polar winds is the immediate cause of Africa's drought. The polar winds block moisture-bearing equatorial winds from bringing rainfall to the parched sub-Sahara region and other drought-ridden areas around the world from Central America to the Middle East and India. The same winds have created different weather quirks in the United States and other temperate zones.

Many of these meteorological and climatic changes are in line with the general picture given by my psychic group. The group said that our weather will continue to become more severe, with temperatures getting even colder. They warned of hurricanes and tornadoes whipping across the earth with increasing frequency, and that these events would also bring pestilence and famine. One climate researcher sounds like he was a member of my psychic group when he asks, "Can the flap of a butterfly's wings in Peru cause a tornado in Iowa?" The researcher, Professor Edward N. Lorenz of the Massachusetts Institute of Technology, is serious, according to an article about our changing climate in *National Geographic* (November, 1976). There Lorenz argues that a random tiny disturbance to the atmosphere on one side of the globe might kick off others that could change an entire weather pattern on the other side of the globe. This concept is quite similar to the "interconnectedness" of the geological changes the psychics foresee. They see the developing severe weather pattern culminating in an entirely new climatic regime after the earth suddenly tumbles on its axis.

Psychics and scientists both see dire consequences for our changing weather and climate. *Science News* notes that Dr. Reid Bryson of the University of Wisconsin has long argued that even slight climatic changes can have devastating effects on crops, wildlife, and people. A 1976 Central Intelligence Agency report draws a grim picture of the future: "The new climatic era brings a promise of famine and starvation to many areas of the world . . . The economic and political impact of major climatic shifts is almost beyond comprehension." The report says the millions in India will face starvation, China will suffer a major famine every five years, and the Soviet Union will lose a major wheat-growing area because of catastrophic changes in the earth's climate. The CIA report indicated that a Northern Hemisphere temperature drop of only 1° C. could precipitate these calamities. The CIA is particularly concerned with the massive migrations of people that would be set in motion by food shortages and the actions a militarily powerful but hungry nation might take.

The question now is: what is behind our changing climate? What is causing the climate to grow colder? Here are two possibilities.

1. A University of California research team believes that man has a hand in turning our planet colder. The team thinks that widespread clearing of tropical jungles is contributing to the drop in temperature. They reason that since the jungles absorb the sun's warming rays and keep the atmosphere moist, clearing the jungles causes the sun's energy to be reflected back into space and causes less recycling of moist air. They warn that their computer simulation experiments show that if current widespread jungle clearing continues, it will cause world average

temperatures to drop about one-third of a degree Celsius. One-third of a degree may seem insignificant, but Iceland is less than a degree and a half away from the statistically normal temperatures that generated the "little ice age" that froze Europe three hundred years ago. In fact, past temperatures of the earth have only been as high as they are now 5 percent of the time. Climatologists say that for fifty of the past sixty years the world's climate has been *abnormal*, not normal.

2. Reid Bryson of the University of Wisconsin and many other climatologists believe that man is turning the climate colder by his farming and fuel-burning activities. These activities, they say, release dust particles into the atmosphere that block sunlight from reaching and heating the surface of the earth. These increased atmospheric particles act like tiny mirrors reflecting back some of the sunlight striking the atmosphere.

In addition to these two possibilities, we should also consider what the effects of some large-scale engineering projects now on the drawing boards will be. The USSR plans to divert some of the great Siberian rivers into vast irrigation projects. These rivers normally empty into the Arctic Ocean, where the lighter fresh water spreads out over the salt water and permits the Arctic seas to freeze. Some experiments indicate that one consequence of preventing the freezing of the Arctic Ocean is likely to be colder winters for the middle latitudes.

Beyond the effects of human activities, some scientists say the climate seems to be turning colder simply because another ice age is coming. This naturally raises the question of what causes ice ages and why one might be approaching at this time.

In the last million years, major glacial periods have advanced and retreated four times. Glaciers covered most of the Northern Hemisphere, reaching as far south as Indiana and Ohio in the United States and as far as France and Rumania in Europe, so that a large portion of land was covered with ice. But as Reid Bryson points out, this only constituted "...a mere 6 percent difference" from today's cover. And during those ice ages, the world's average temperature was only seven to ten degrees colder than today. The last maximum ice cover was about 16,000 B.C. and it wasn't until 3000 B.C. that the ice was confined to Greenland and Antarctica, as it is today.

What brings on the ice and cold? There are as many as sixty different, often conflicting theories to explain climatic changes, including variations in solar radiation, the earth's magnetic poles, ocean circulation, and the amount of volcanic dust in the stratosphere. Most experts agree, though, that the marked increase of coal and petroleum products during the first half of this century—via the so-called greenhouse effect—has raised the earth's temperature slightly during this period, masking some of the overall cooling that is now taking place.

One theory, however, has been generally accepted as the "fundamental cause" of ice ages. It states that glacial periods are brought about by variations in the earth's movement around the sun. This concept was originally championed by Alfred Wegener, the same man whose ideas about continental drift have recently been proven correct. Then in the 1930s this climatic theory was refined by Yugoslav mathematician Milutin Milankovitch, for whom the theory is now named.

The Milankovitch theory could not be tested until several years ago when it became possible to get an accurate record of the earth's history. In an unusually positive statement, geologist Dr. James D. Hays of Columbia University's Lamont-Doherty Geological Laboratory told *Science News*, "We are certain now that changes in the earth's orbital geometry caused the ice ages. The evidence is so strong that other explanations must now be discarded or modified."

Working with Dr. Hays were Dr. John Imbrie, a professor of oceanography at Brown University, and Dr. Nicholas J. Shackelton, a geologist at Cambridge University in England. Their research was sponsored by the National Science Foundation as part of a project called CLIMAP (Climate Long Range Mapping and Prediction). The group described their research in detail in the December, 1976, issue of *Science*, where they stated, "It is concluded that changes in the earth's orbital geometry are the fundamental cause of the succession of Quaternary ice ages."

The group based their conclusions on an analysis of microorganisms preserved in sediments beneath the floor of the South Indian Ocean. Deep-sea cores or bottom samples (now obtainable as a result of recent technological advances) from all parts of the world provide the most important source of past climatic data. The Indian Ocean cores give an uninterrupted record of the earth's climate dating back 450,000 years—three times further than scientists had been able to go before. The cores contain a record of all climatic reversals during this period. These climatic reversals were determined by analyzing the alternating layers of species of warm- and cold-preferring microfossils known as radiolaria contained in the core. Because radiolaria are particularly sensitive to temperature changes, by analyzing the relative abundance of those preferring warmth or cold, an estimate of the prevailing water temperature during their lives can be made.

Further evidence came from measuring the ratio of two different types of oxygen molecules in the shells of the radiolaria. This enabled the scientists to determine past changes in global ice volume—that is, the amount of water bound in polar ice caps at a given time. This independent dating of ice ages coincided very closely with that inferred from the radiolaria data. *Science News* explained this second technique thus:

> As water molecules evaporate from the sea surface, those containing an atom of oxygen-16 isotope will tend to leave the surface faster than the heavier molecules containing oxygen-18. During the ice ages, the oxygen-16 molecules are thus more likely to be bound up in the polar caps when they fall back to earth in snow. An enrichment of oxygen-16 molecules in a sediment [core] layer therefore indicates an ice age.

* * *

So from the cores the research team had a record against which they could compare Milankovitch's calculated climatic cycles of 93,000, 41,000, and 23,000 years. The record preserved in the sea cores indeed showed these same cycles, which allowed the group to declare so positively that the Milankovitch theory had been confirmed. The timing of each of the major ice ages was closely related to the known cycle of changes in the earth's attitude (orbital tilt) and orbit, changes that reduced or increased the amount of sunlight striking the polar caps. Scientists believe the changes in the earth's orbital geometry are caused by the gravitational tug of the other planets, particularly Jupiter.

The longest cycle measured in the cores—about ninety-three thousand years—matched the length of cyclical changes in the shape of the earth's orbit, ranging from nearly circular to elliptical and back. The earth's orbit is now becoming more eccentric, which increases seasonal differences in the distance between the sun and the earth. This in turn tends to increase the severity of ice ages.

The length of the second cycle in the cores—about forty-one thousand years—matched the length of changes in the tilt of the earth's axis of rotation with respect to the earth's orbit around the sun. This tilt varies from 22 to 24.5 degrees from perpendicular over the forty-one-thousand-year period. If there were no tilt and the earth's axis of rotation were always perpendicular to the direction of the sun, there would be no seasons and polar ice would never melt. The earth's trend now is toward minimal tilt, which foreshadows a new ice age.

The length of the shortest cycle in the cores—twenty-three thousand years—corresponds approximately to the precession of the equinoxes. In this cycle the earth advances in its elliptical orbit so that its closest approach to the sun occurs at different times of the year. When the distance between the earth and the sun is the greatest in the summer, it results in cooler temperatures, less snow melting, and a growth of the polar ice caps. The earth presently is in this position and it is leading us into a new ice age.

Based on the relationship between the ice ages and the earth's positions in space, Dr. Hays and his group of scientists believe that the earth is moving toward "extensive Northern Hemisphere glaciation and cooler climates." Recall the drop in temperature now occurring, the increasing snow and ice cover, the cooling North Atlantic, and the expanding polar winds noted earlier. "As the cooling trend continues," Dr. Hays told the *Los Angeles Times,* "it will accelerate when the ice begins to accumulate, and that could take place maybe even in the next thousand years."

This echoes Bryson's statement to *Science News* that "... ice ages come on rather abruptly ..." An example of this comes from the *National Geographic* article cited earlier that reports that "one drop from fairly warm times into full ice age cold took place in Greenland 89,500 years ago—apparently in less than a hundred years." In 1971 a geophysical team working at Camp Century, Greenland, discovered evidence of this abrupt climatic change in ice cores from the Greenland ice cap. The team said that evidence of this change was supported by drastic changes in plant life along the Gulf of Mexico. In a colloquium on their findings discussed in *The Age of Cataclysm*, the team reported that "the conditions for a catastrophic event are present today."

In general, psychics have not said much about climate. I offer this chapter because it concerns an earth-change factor obviously taking place at this moment, and I am not limiting myself only to what psychics have predicted. However, to repeat, my psychic group observed that the forces acting within and upon the planet are interconnected. Changes in climate are related to the likelihood of earthquakes via the lowered water table, and both are related to processes occurring deep inside the planet and in outer space (as Chapter 8 will show).

There is a further connection between climatic changes and impending disaster through earth changes that will become clear in the next chapter, when we examine predictions of an axis displacement or pole shift. To anticipate: a pole shift may result from either a significant change in position or a change in mass of one or both of the polar ice caps. Such an event could affect the planet's gyroscopic stability to the point where recovery of balance is impossible, and a tumble in space would occur. Thus, if a new ice age is coming upon us, with a buildup of the polar caps, it is just possible that the stage is being set for the greatest of all cataclysms, short of total destruction of the planet.

CHAPTER 7

Bermuda Shorts in Alaska, or the Shifting of the Poles

> "Therefore I shall shake the heavens and the earth shall be moved out of its place..."
>
> Isaiah 13:13

> "...the doctrine of 'uniformitarianism'...has governed scientific thinking...It is so entrenched, in fact, that the revival of dramatic explanations for some developments in the earth's history has been greeted with great hostility."
>
> Walter Sullivan
> *Continents in Motion*, 1974

There may be something more behind the occurrence of ice ages than changes in the earth's orbit, as discussed in the previous chapter, something even more critical. Some sort of castastrophic activity comes to mind.

Louis Agassiz, the father of glacial geology and the first man to recognize (in 1840) that distinct ice ages occurred, held that the beginning of an ice age was a global catastrophic event accompanied by a sudden drop in temperature. Agassiz's research and ideas are all accepted by geology to this day except his ideas about the catastrophic start of ice ages; these have been recast to accord with the ruling principle of uniformitarianism. Georges Cuvier, the father of paleontology, and Louis Agassiz, the father of glacial geology, were both catastrophists. Considering the vindication of Alfred Wegener on both continental drift and the earth's orbit playing a key role in the climatic changes, one may wonder if the other theories of geology's pioneers are also correct. Have modern geologists blinded themselves to the truth?

Unfortunately, Agassiz never discussed a mechanism that could bring about the catastrophic onset of ice ages. But in 1911, an electrical engineer named Hugh Auchincloss Brown, interested by reports of mammoths found frozen in the Arctic with "buttercups still clenched between their teeth," began framing a theory that the accumulation of ice at one or both poles periodically upsets the equilibrium of the spinning earth, causing it to tumble in space "like an overloaded canoe." This would cause catastrophic worldwide climatic changes, including the start of a new ice age, with the oceans sweeping across the continents in global floods, extinguishing many life forms. Such an event, he said, would indeed "quick-freeze" the mammoths found buried in a standing position in the Arctic.

The earth has a bulge along the equator. The circumference of the earth is twenty-seven miles greater around the equator than around the poles. As an engineer, Brown knew that this equatorial bulge stablized the earth's spin. Despite this stablizing effect, Brown reasoned, a large accumulation of ice in Antarctica would be enough to unbalance and topple the earth, which he pictured as a spinning top. Brown argued that as a result of these ice accumulations the earth periodically capsized at intervals of about eight thousand years, with each catastrophe nearly obliterating civilization as it existed at that time. When Brown died in 1975 at the age of ninety-six, the ice in Antarctica was two to three miles thick and still growing. Brown believed that the next capsizing was long overdue. His ideas were recently dramatized in a novel, *The HAB Theory*, by Allen W. Eckert.

Although Brown himself has been largely ignored, the theme of global capsizing was taken up seriously in 1964 by members of Princeton's Department of Geology when they debated a proposal by A. T. Wilson of Victoria University in New Zealand. Wilson theorized that ice ages might have been rapidly brought on by periodic catastrophic surges of the Antarctic ice sheet. In his view, large surges of ice comparable to those observed in the last century would dramatically increase the reflecting surface of the South Pole. For example, between 1935 and 1938 part of the Spitsbergen ice cap advanced more than thirteen miles on a front twenty miles wide, and the Kutiah Glacier of northern India is said to have advanced as much as 360 feet a day at times. As a result of ice surges, more sunlight would be reflected back into space so that the entire atmosphere of the earth would be cooled, thereby starting an ice age. Such an occurrence would provide a sudden additional ice accumulation to help in unbalancing the earth and capsizing it, as Brown feared.

The major clue predicting an actual tumble in space, according to Brown, is the wobble in the earth's axis of rotation, or spin axis. As the earth spins about its axis, the axis itself gently rocks or wobbles in space. This wobble is still a puzzle to scientists. The earth's spin axis is affected by two factors. The first is external forces such as the gravity of the moon and the sun acting on the earth's equatorial bulge. The second is internal forces such as changes in the distribution of the earth's mass. If the earth's mass were uniformly and symmetrically distributed about its axis, there would be no wobble and the earth's spin would be stable. But the distribution of the earth's continents, oceans, and ice is not even. In fact, this mass distribution is constantly changing as land erodes, continents drift, and the ice caps grow or shrink in size. Brown said that an accumulation of ice or snow at either pole increased the wobble to the point where the spin would become unstable and the earth would eventually tumble in space.

While the scientific establishment has ignored Brown's thesis, some scientists have independently pointed to similar possibilities. In 1952 Drs. Walter Munk and John Revelle, both of the Scripps Institute of Oceanography at La Jolla, California, described how large amounts of ice melting sufficiently fast (Brown's proposition in reverse) could affect the earth's rotational axis and slow the earth's spin. The men cited evidence from celestial observations that the earth's rotational rate had slowed and that its axis had abruptly shifted three times in the last hundred years. The shifts resulted in a change of about .006 of a second in the length of the day. Though these shifts were very minute, they are impressive if one considers the mass of the earth. In another paper these same researchers proposed that the cause of such sudden shifts may lie deep in the earth's interior and be reflected on the earth's surface by rapid drift of the magnetic field. One advantage of their new theory, they said—taking tongue-in-cheek notice of Brown's theory—is that the core of the earth "is even less accessible than the Antarctic."

In 1960, Dr. Thomas Gold, then at the Royal Greenwich Observatory, wrote in *Nature* that if the earth's interior is plastic, the earth's equatorial bulge is not that great a stabilizer, or help in preventing capsizing. Indeed, it is because of this plasticity that the bulge exists as a by-product of the earth's spin. Reminiscent of Brown, Gold suggested that a drift in the spin axis could be induced by accumulations of matter, such as the rapid building up of an ice sheet. Walter Sullivan, paraphrasing Gold's argument in *Continents in Motion*, said ". . . if a mass of ice accumulated in a continent midway between the equator and one of the poles, this would exert a substantial disruption effect on the spin. The axis would slowly change as the equatorial bulge adjusted toward the new spin axis."

Gold said polar shifts have apparently occurred at intervals of millions of years in the geologic past and have brought about great changes in climate and evolution of species. Illustrating what a large change in the earth's distribution of mass could do, Gold wrote in 1955 in *Nature*, "If a continent of the size of South America were suddenly raised by 30 meters . . . the earth would hence topple over . . . It is thus tempting to suggest that there have been just a few occasions when the earth's axis has been 'free' and has swung around." Thus the only difference between Gold's argument and Brown's is (1) the frequency and amount of axis shift and (2) how fast such an axis shift takes place.

Even though Gold is part of the establishment, no one has yet taken serious notice of the possibility of the earth periodically capsizing. Not surprisingly, the clever Alfred Wegener once put forth a theory very similar to Gold's. Wegener's theory also involves an adjustment in equatorial bulge. Like Brown, however, Wegener believed that the change in the earth's axis took place quickly—indeed, fast enough to cause extensive flooding through the oceans' inundating equatorial lands. Intriguingly, in 1971 and 1972 the close-up pictures of Mars sent to earth by Mariner 9 revealed a series of concentric features around both Martian poles. Drs. Bruce Murray and Michael Marlin of the California Institute of Technology, pointing out that these features are not centered on Mars's present poles, propose that the features are evidence of past changes in the Martian spin axis.

In 1958, two men outside the geologic establishment took up Brown's catastrophism theory. Charles H. Hapgood, a professor of history and anthropology at Keene State College in New Hampshire, and James H. Campbell, an engineer who had helped develop the Sperry gyroscopic compass (and was therefore familiar with the laws governing spin stability), decided that instead of the earth as a whole tumbling over during a pole shift, only the earth's crust moved. The planet's rigid mantle, they theorized, slides over the earth's molten interior like the loose skin of an orange slipping over the fruit inside. The global network of deep rifts and ridges around the world, compared by them to the cracked shell of a hard-boiled egg, was taken as evidence for such crustal sliding. They pointed out that the huge sheet of ice covering Antarctica was not very well centered on the South Pole and calculated that its center of gravity lay more than 345 miles from the polar axis. As a result, this huge ice mass is rotating eccentrically, much like a rug in a spinning washing machine. It must be remembered that Antarctica is nearly as big as North America. It has mountain ranges comparable in size to the Rockies, covered with ice from one to three miles thick. (Note that whereas the Antarctic is under ice, the Arctic except for Greenland is covered with ocean and ice floe.) Hapgood and Campbell argued that ice periodically accumulates at one of the poles so fast that the earth's spin axis can't adjust for the off-centered mass and so the earth's crust slips to compensate. The crust shifts so much, they say, that great changes in the positioning of the poles are produced.

None other than Albert Einstein wrote the introduction to Hapgood and Campbell's book, *Earth's Shifting Crust*. Einstein stated:

> I frequently receive communication from people who wish to consult me concerning their unpublished ideas. It goes without saying that these ideas are very seldom possessed by scientific validity. The very first communication, however, that I received from Mr. Hapgood electrified me. His idea is original, of great simplicity, and—if it continues to prove itself—of great importance to everything that is related to the history of the earth's surface.

Einstein corresponded extensively with Hapgood and made a variety of suggestions and comments. Einstein even proposed that Hapgood be appointed to the Institute for Advanced Study in Princeton, and to a Guggenheim fellowship—though both proposals were promptly turned down. It seems that no one (besides Einstein) was convinced that an increase in the ice mass alone could cause the earth's crust to slip and/or the earth to suddenly tumble over. A better or additional mechanism was needed.

Just such a mechanism was discovered a few years ago, but scientists, having little interest in researching catastrophic events, failed to make the connection between this discovery and the onset of ice ages. It is known that at certain times the earth's wobble increases and polar drift reaches a maximum, but that the internal properties of the earth gradually reduce or dampen this heightened activity. Scientists have long puzzled over what produces this occasional heightened activity. It is believed that the solution to this problem is linked to whatever provides the energy to maintain the earth's more normal wobble. Some scientists suspected that great earthquakes gave the earth's spin a jolt from time to time, and were responsible for these periods of heightened activity.

Beginning in 1955, new and more careful observations of the earth's spin relating to wobble were made by many different astronomical observatories under the auspices of the international time service (the Bureau International de l'Heure—BIH) in Paris. For the first time wobble measurements were reduced to plot the exact motion or path of the North Pole. By early 1967, thirty-nine different observatories were contributing daily information on pole positions. Drs. E. Smylie and L. Mansinha of the University of Western Ontario in Canada studied the new data on pole movement. In September, 1968, they announced that the North Pole—that is, the north end of the spin axis—followed a counterclockwise circular path (actually a gentle inward spiral). But around the time of large earthquakes something remarkable happened. The North Pole actually shifted its position while equal and opposite force was contributed to the earth's wobble. Thus, the position of the pole was left nearly unaltered; only its circular path changed somewhat.

Determinations made by the BIH showed that temporary changes in the pole path closely correlated with the occurrence of fifteen of the total of twenty-two earthquakes with a magnitude greater than 7.5 that occurred in the period examined, 1957–68. For the other seven quakes, Smylie and Mansinha emphasized that there was evidence of a systematic deviation in the path, although the deviation was not revealed by the circular arcs described by the most clear-cut deviations. Thus it seems that very large earthquakes might cause the earth to tumble in space and quickly bring on an ice age like the one that swept across the world in less than a century 89,500 years ago.

What might a *series* of earthquakes do? For example, Roman records report that in just one year during the Punic Wars (217 B.C.) fifty-seven large earthquakes were reported in Rome. And Massachusetts Institute of Technology scientists have proposed that every fifty years, a quake four times more severe than the Alaskan Good Friday quake (8.3) occurs somewhere on the earth. *If a 7.5–8.5 magnitude earthquake is related to a clear-cut deviation in wobble, what could a 10.0 or 12.0 magnitude earthquake do if it struck when the earth's balance was already compromised by an unusually large ice accumulation at either of the poles?* (While pole path and wobble change several days before the actual quake, for practical purposes we can consider the quake to be the cause since it is the final output of this as yet unknown process that produces earthquakes. The earthquake provides the final jolt, so to speak.)

All the psychics I worked with talked about the earth periodically tumbling on its axis, and they all foresee this happening again around 2000 A.D., and it is not likely that they were aware of the scientific speculation since such papers are read only at special conferences and published in only obscure scientific journals. The pattern of increasing natural disaster—from earthquakes and volcanoes to land rising and sinking and to freakish weather patterns—all was foreseen as culminating in this greatest of global events, the pole shift. Some psychics foresaw only a 60-to-90 degree tumble; some foresaw even greater movement before the earth rebalanced itself and resumed its path around the sun. One jokingly suggested that people in Alaska should buy Bermuda shorts. More seriously, all suggested that the human race make itself ready for the drastic climatic changes that would follow if the earth suddenly lurched on its axis. Cayce also foretold what we could expect at the end of this century: ". . . there will be the shifting then of the poles—so that where there have been those of a frigid or semi-tropical will become

DISPLACEMENT OF POLE PATH AS A RESULT OF LARGE EARTHQUAKES

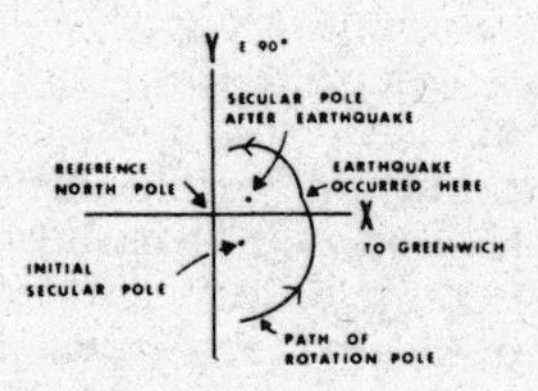

Effect of an earthquake on the pole path. Displacement field is assumed to be established suddenly.

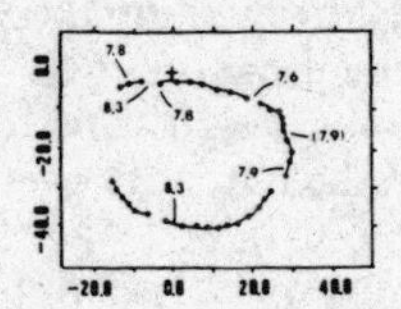

Pole motion anti-clockwise. Determinations of the Bureau International de l'Heure (BIH). All earthquakes of magnitude M>7.5 are indicated. Note only one earthquake (in parenthesis) did not result in the pole path being displaced.

adapted from Mansinha, L. and Smylie, D.E. – Science Sept. 1968

Disruption of pole path as a result of large earthquakes.

more tropical and moss and fern will grow" (396-15, January 19, 1934).

While earth scientists have for the most part ignored the possibility of global catastrophe due to the earth tumbling on its axis, this has not daunted Velikovsky, who has combed vast amounts of ancient literature and claims to have found evidence of people actually witnessing such sudden axis shifts. In *Worlds in Collision* (1950), Velikovsky cites accounts in Sumerian, Chaldean, Hindu, Chinese, Mayan, Aztec, Icelandic, Egyptian, and Hebrew records—accounts that show a striking correspondence. For example, he notes that Pomponius Mela, a Latin author of the first century, wrote: "The Egyptians pride themselves on being the most ancient people in the world. In their authentic annals . . . one may read that since they have been in existence, the course of the stars has changed direction four times, and that the sun has set twice in that part of the sky where it rises today." Velikovsky observes that the Greek historian Herodotus, writing in the fifth century B.C., makes the same points as Pomponius does, which has puzzled his commentators. Velikovsky also shows how the *Magical Papyrus Harris* speaks of a cosmic upheaval of fire and water when "the south becomes the north and the earth turns over." Various panels in early Egyptian tombs, Velikovsky points out, show the celestial sphere in a reversed orientation, where north and south and east and west have been interchanged.

Velikovsky challenges the natural sciences to resolve the historical and literary evidence he presents indicating axis shift. He says that when the poles flip it takes the earth several days to start up its normal spin again. In such a situation, he says, the sun and the planets would appear to be locked in place until diurnal spin begins again. From the Book of Joshua and the Midrash, in the collection of ancient Jewish discussions of Scriptural passages, Velikovsky notes statements telling how the sun and moon stood still for a period of time. From the Mexican annals of Cuauhtitlán he notes the discussion of a cosmic catastrophe that occurred in the remote past where the night did not end for four days. The *Manuscript Quiche* of the Mayas tells that in the Western Hemisphere in the days of a great cataclysm the sun's motion was interrupted. Velikovsky tells of Sahagun, a Spanish soldier who came to America in the early 1500s, gathered Indian traditions, and recorded that at the time of one cosmic catastrophe the sun rose only a little way over the horizon and remained there without moving and that the moon also stood still. In short, Velikovsky shows that ancient documents and traditions make it possible to infer an event of global proportions in which, at various points around the world, there was either prolonged night, prolonged day, or the sun or moon tarried at different points in the sky.

While the geological establishment attacked Velikovsky for his heretical thesis, some members of the press were truly impressed with what Velikovsky had to say. As long ago as 1946 the late John J. O'Neill, science editor of the now defunct *New York Herald Tribune*, said Velikovsky's thesis was derived from "a magnificent piece of scholarly historical research." He described it as "a stupendous panorama of terrestrial and human history which will stand as a challenge to scientists. . . ." Over the years many of Velikovsky's ideas about the physical characteristics of some of the planets in the solar system have been shown to be amazingly accurate. Walter Sullivan says in *Continents in Motion* that "Velikovsky, by any mode of measurement, is an extraordinary man."

One flaw in Velikovsky's argument is that he regards most of the accounts he cites as firsthand accounts, which cause him to believe that pole shifts have occurred frequently and recently—the last only five thousand to eight thousand years ago. But there is little evidence in the geological record to support such recent pole shifts. So it is easy to understand why many geologists ignored Velikovsky's arguments, along with Brown's. Unfortunately, no geologist was concerned enough to consider the possibility that these accounts may have been passed down orally through the ages, and were written down only centuries after the events occurred. Since Velikovsky wrote, Alexander Marshack, a Research Fellow at the Peabody Museum of Archaeology and Ethnology of Harvard University, has shown that seemingly meaningless markings on pieces of bone from Europe dating from thirty thousand to seventy thousand years ago correlate with phases of the moon. Thus it seems that man was perfectly capable of appreciating and recording celestial phenomena at very early dates—far earlier than the establishment now recognizes. Most recently, a similarly marked piece of bone has been found at Pech de Laze in France, a 300,000-year-old site.

My own ongoing archaeological excavation at Flagstaff, Arizona—detailed in *Psychic Archaeology*—is in the area where the Hopi say their previous "worlds"—highly developed civilizations—existed. The tribe's tradition teaches that three worlds existed prior to the one in which we now live. In *The Book of the Hopi*, Frank Waters points out that Hopi legend maintains that at the close of their second world, the twin gods responsible for keeping the earth rotating properly abandoned their stations. "Then the world, with no one to control it, teetered off balance, spun around crazily, then rolled over twice. Mountains plunged into the sea with a great splash, seas and lakes sloshed over the land; and as the world spun through cold and lifeless space it froze into solid ice." I am eager to see if my deeper excavation will in some way confirm this legend.

Besides their second world being destroyed by ice, the Hopi say that their first world was destroyed by fire and their third world by water. This all makes good geological sense. The destruction of the first world by fire could represent the volcanic activity that took place in the San Francisco Mountains outside of Flagstaff over two hundred and fifty thousand years ago. The destruction of the second world by ice could represent the glacierlike activity that took place in the peaks approximately one hundred thousand years ago. And the destruction of the third world by water could represent the wet period and corresponding inner-mountain basin damming and flooding that occurred approximately twenty-five thousand years ago. The main problem here is that this would establish the presence of humans in the New World two hundred and fifty thousand years ago—twenty-five times earlier than the generally accepted date. The orthodox view holds that approximately ten thousand years ago the American Indians' ancestors entered the New World from Asia by walking across a land bridge that then existed in the Bering Strait. However, my excavation at Flagstaff has already shown that man was here more than one hundred thousand years ago.

We can approach the possibility of pole shift from still another direction. The earth occasionally tumbling on its axis explains more than only the onset of ice ages, myths, and frozen mammoths. It also explains a great number of other correlations and anomalies found in the earth's past. In fact, *sudden pole shift is a unifying explanation for mountain building, volcanic activity, continental drift, magnetic pole reversals, animal extinction, and the erratic occurrence of glaciers.*

Geologists find it difficult to explain why, in the distant geological past, large ice accumulations occurred in what are now tropical and semitropical regions. Extensive ice sheets once existed in South America, Australia, Africa, and India. By examining the deposits left by these glaciers and the trend of the scratches and grooves etched into bedrock, geologists have also found that these anomalous glacial sheets moved in directions opposite to what would be expected. Dr. William Stokes of the University of Utah, in his text *Essentials of Earth History*, points out,

> In South Africa the glaciers moved principally from north to south—away from the Equator. In central Africa and Madagascar other deposits show that the ice moved northward, well within the tropic zone. Most surprising has been the discovery of great beds of glacial debris in northern India, where the direction of movement was northward . . . in Australia and Tasmania, where the ice moved from south to north . . . movement in Brazil and Argentina was toward the west.

Dr. C. O. Dunbar of Yale was awed by how the glaciation in Brazil occurs within 10 degrees of the equator and how in India the ice flowed "from the tropics to the higher latitudes." Many geologists believe that the north and south poles once were located in these now warm regions, but only because of continental drift. Continental drift is too slow a process to account for the amount of movement involved and it doesn't explain all the locations observed. Pole shift seems the most viable way of changing polar location. The anomalous occurrence of coal formations and coral in present-day polar regions seems to be another manifestation of the once different locations of the poles. Coal and coral only develop in tropical areas, yet both are found below the snow of Spitsbergen in the Arctic Ocean (78°56′ north latitude), only slightly more than 9 degrees from the North Pole, and coal seams have also been found in Antarctica. Dr. W. B. Wright in *The Quaternary Ice Age* says that during geological history many changes in the position of the earth's climatic zones occurred that cannot be explained except by a displacement of the earth's axis. Wright says that since such displacement explains one of the oldest glacial ages, "... it becomes worthwhile to inquire whether the Quaternary [recent] glaciation would not have a similar cause."

I once asked Aron Abrahamsen for the location of the poles during different time periods of the past. I was impressed with the way the locations he gave matched up with the different anomalous glacial locations and how the time periods he gave corresponded with the dates of the earth's last major glaciations. One would think Abrahamsen was studying geology books instead of the Bible, which he does daily. Cayce also talked about several pole shifts in the past and indirectly gave the locations of these past polar positions. This information greatly impressed the distinguished geologist who studied his work but who, fearing for his career, chose to remain anonymous.

Pole shift may also explain the overwhelming association between ice ages and mountain building. Stokes, in

Erratic ice deposits and movements indicating possible former pole positions

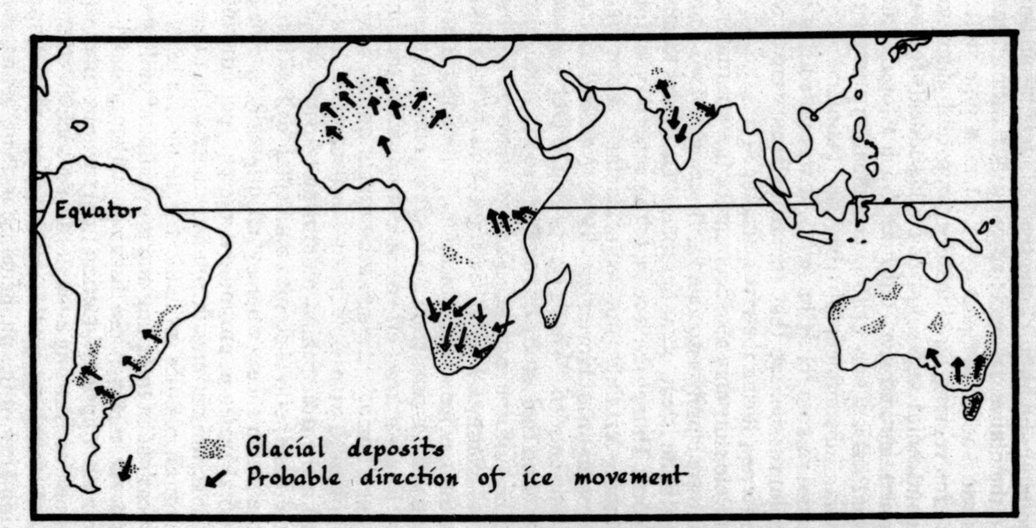

adapted from Gilluly

his *Essentials of Earth History*, says that "the one invariable accompaniment of glaciation appears to be mountain-building. Intensive mountain-building and continental elevation have coincided with all the great glaciations of the past..." Just as a pole shift would suddenly bring on climatic change and glaciation, the tremendous forces involved could suddenly force up new mountains and level old ones. Such a force could have propelled a slowly drifting continental mass such as the Indian subcontinent into the Asian mainland to force the Himalayas up. (In Chapter 3, it was noted that certain geologists believe that the Himalayas were driven up suddenly during the last glacial age.)

Pole shift may also account for the volcanic eruptions that are closely associated with ice ages. Two University of Rhode Island oceanographers, Dr. James Kennett and Robert Thunell, believe that "this general synchronism is almost certainly not coincidental." Their remarks, reported in a November, 1976, *National Geographic* article entitled "What's Happening to Our Climate?" are based on their study of volcanic ash in seabed cores. The author of the article, *National Geographic* assistant editor Samuel W. Mathews, asks how "volcanoes trigger ice ages, or ice sets off volcanoes." This uncertainty can be resolved. A pole shift could create enough stress in the earth's crust to set off volcanic eruptions at the same time as new climatic regimes are brought about.

Finally, pole shift is an excellent candidate for the driving force behind continental drift. While most geologists now accept continental drift, there is no convincing theory about what originally set the earth's crustal plates in motion. At present, periodic intensive convection currents or plume activity within the earth is the most popular candidate for the motive force. Others are sea-floor spreading and gravitational pull by the sun and the moon (tidal tugging). Other geologists have pointed out that neither tidal tugging nor sea-floor spreading is strong enough to have such an effect on the earth's drifting plates. Others note that despite the popularity of the plume concept, there is no convincing evidence for their existence. On the other hand, pole shift could suddenly propel the continents forward. The slow continental drift observed today may be the residual momentum of a once great surge.

Alfred Wegener may have again been prophetic when he said that changes in the earth's spin axis from time to time would bring about new directions of continental drift. We have seen that changes in the earth's spin axis could arise from polar shifting. Wegener pointed to the present westerly movement of the continents as evidence of this spin effect. In 1972 Drs. Leon Knopoff and A. Leed of the University of California at Los Angeles analyzed in *Science* the motion of the ten major crustal plates and found an overall present tendency to westerly drift. Critics note that in the past the movements of the continents were northerly or southerly and unrelated to present spin. But if polar shift is a reality, these previous directions of movement could relate to past pole locations and past spin directions, which polar shift could easily bring about. For example, as noted earlier, geologists point to a sudden change of direction in the 3,700-mile volcanic Hawaii–Emperor seamount chain as evidence for drift, the crustal plate slipping over a stationary hot spot. But they don't say what caused the sudden, approximate 110-degree change in the direction of drift. Pole shift may again hold the answer.

One of the most important anomalies that pole shift can explain is the periodic reversal of the earth's magnetic axis. The behavior of a compass indicates that the earth is a gigantic magnet with positive (north) and negative (south) poles that lie near the geographic poles. Although magnetized material lies deep within the earth, the exact cause of the earth's magnetism is not fully understood. According to Drs. Allan Cox, Brent Dalrymple, and Richard Doell, three of the USGS's top scientists, "After centuries of research the earth's magnetic field remains one of the best described and least understood of all planetary phenomena." Beyond slow and progressive polar wandering, scientists do not understand why the poles have alternately changed their magnetic polarity over the ages. Why has the North Pole, for example, alternately been positive and negative? In the past 76 million years the north and south poles have changed magnetic polarity at least 171 times. Within the past 48 million years, polar magnetic records recorded in rocks and sediments show that there have been about five reversals per million years, with the average time between reversals about 220,000 years and the shortest time about 30,000 years. But in the last one million years there has only been one reversal, approximately 700,000 years ago. Applying the principle of uniformitarianism, we can surmise that another reversal is long overdue. In fact, over

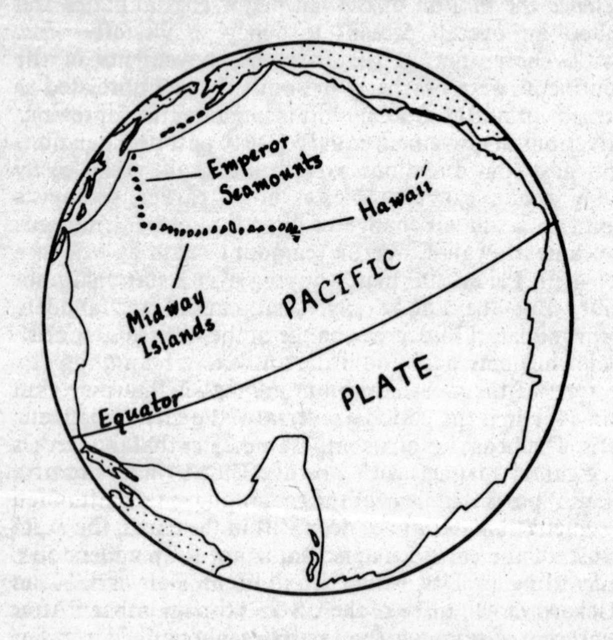

Inexplicable change in direction of Hawaiian volcanic chain.

the past twenty-five hundred years the total magnetic field of the earth has weakened by about 50 percent and some experts feel this may be a prelude to a new reversal within a few centuries.

In addition to the magnetic "long-term" reversals noted, there are "short-term" reversals of only several thousand years. "Short-term" magnetic changes not only involve full 180-degree reversals but also magnetic deviations of 30 to 90 degrees. Full 180-degree magnetic field reversals have been interpreted as simple electromagnetic field changes rather than 180-degree changes of the rotational axis. In other words, magnetic reversals are believed to be only the result of internal field changes rather than of the earth tumbling over in space. But geologists have not yet found any convincing reason why the earth's magnetism should spontaneously reverse in polarity from time to time. On the other hand, my psychic group emphasized the physical reality of the pole changes they foresaw by describing violent winds (up to 200 miles per hour) that would result from a physical or geographical pole shift rather than simply a magnetic reversal caused by internal reasons.

While the geological establishment rejects rotational pole shift as too catastrophic, it seems no more "far out" a possible cause for the earth's magnetic reversals than Dr. Bruce Heezen of Columbia University's Lamont–Doherty Geology Laboratory's speculation that reversals may occur when the earth is struck by giant meteorites or comet heads. In *Scientific American* (1967) Heezen and his associate Dr. Bill Glass state, "A meteorite large enough to produce a small impact crater falls every few thousand years" and they add, "A cosmic body sufficiently large to cause a reversal of the geomagnetic field may arrive every few hundred thousands of years." They also go on to talk about how "really big impacts" that mark turning points in the earth's history occur "when even larger bodies hit the earth." Heezen and Glass base their argument on the discovery that some sort of extraordinary event spread tiny glass fragments known as tektites over a large part of the earth at about the time of

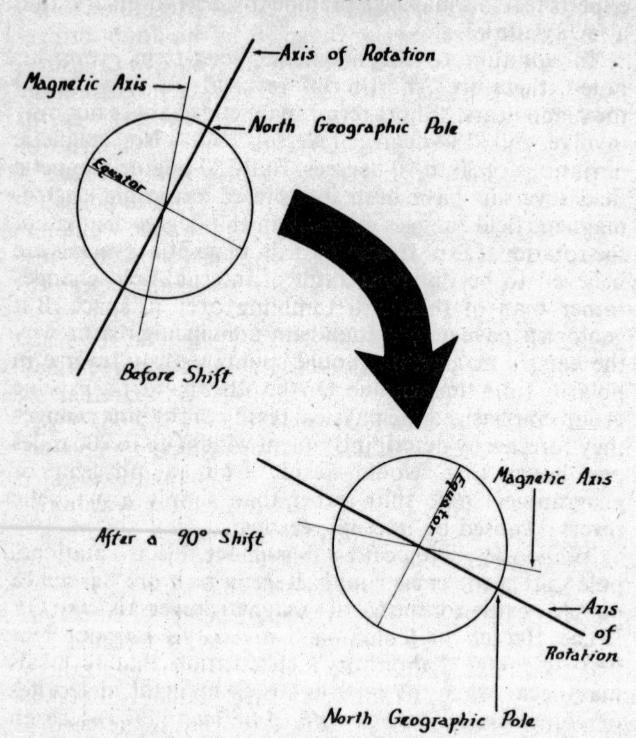

Shifting of the geographical poles, which will result in major climatic changes.

the last major magnetic reversal approximately seven hundred thousand years ago. Many experts believe that these tektites represent splashes of molten rock thrown up by some cataclysmic impact. Even Walter Sullivan, who espouses the establishment point of view in his *Continents in Motion*, admits that "the event responsible for this [tektite] debris must have been fearsome indeed." Other tektite-strewn fields from even earlier periods of time have been found in other parts of the world. The total weight of one fall has been estimated at 110 billion tons. Such talk from establishment scientists no doubt brings joy to the heart of Velikovsky, who in *Worlds in Collision* argued that the earth has gone through a succession of encounters with comets and planets, with the earth tumbling on its axis many times as a result. If the type of collision Heezen proposes occurred when the earth's balance was already compromised due to an unusually large ice accumulation at one of the poles, one wouldn't need a large earthquake to initiate capsizing. It seems likely that the collision would be more than enough to effect a pole shift.

One catastrophe associated with magnetic reversals has recently been well documented—the sudden extinction of animals. Dr. N. D. Watkins and H. G. Goddell of Florida State University stated in *Science* in 1967, "We must consider the possibility of a direct connection between geomagnetic polarity and faunal changes." Dr. James Hazen of Columbia's Lamont-Doherty Laboratory says that "the correlation between reversal and extinction level is indeed striking." He says that "not only did these times of extinction affect a wide variety of animals but they were world-wide in extent," affecting both marine and land animals. Hazen formulated his views from the study of sea cores where he and others found that six species of radiolaria suddenly became extinct during or shortly after switches in the earth's magnetic poles. Such biological havoc also seems related to the occurrence of flash-frozen mammoths in Siberia. Some researchers theorize that deterioration of the earth's protective magnetic shield during reversals allowed an increased amount of damaging cosmic radiation to reach the earth, which radiation killed the animals. But others point out that radiation would not be increased by more than 10 percent, too small an increase to affect life appreciably. Writing in *Nature* in 1971, Dr. S. A. Durrant and H. A. Khan of the Department of Physics at the University of Birmingham, England, correlated various tektite deposits with both magnetic-field reversals and marine-organism extinctions. Likewise in *Nature* in 1970, Dr. J. P. Kennet and N. D. Watkins of the University of Rhode Island Graduate School of Oceanography drew attention to correlations between magnetic field reversals, widespread faunal extinction, climatic changes, and maxima of volcanic activity.

In his books, Velikovsky argues that pole shift has occurred much more recently in time than the last major "long-term" reversal, which occurred 700,000 years ago. One begins to wonder, inasmuch as it wasn't until the last few years that evidence for at least "short-term" reversals and deviations, of less than 180 degrees, which have occurred fairly recently, has been found. From Lake Biwa in Japan, evidence points to short-term magnetic events of 295,000, 180,000, and 110,000 years ago. From Czechoslovakia and Lake Mungo, Australia, there is evidence for a magnetic event 30,000 years ago, from France an event 20,000 years ago, and from Gothenburg, Sweden, an event only 12,600 years ago. These shorter-term events are hard to detect and it is now accepted that many short-term events probably remain to be discovered. But this doesn't explain what causes a 60- or 90-degree "short-term" pole deviation. Hapgood and Brown would say that such a deviation would come from the ice-laden pole dropping down to an equatorial position as a new point of rotational stability is found. While establishment scientists argue for internally derived full 180-degree reversals, these 60- to 90-degree shifts remain a complete puzzle.

Cuvier challenged his geological colleagues in the last century to explain the sudden and tremendous changes he found in the geological record. He said, ". . . it is in vain that we search among the powers which now act at the surface of the earth for causes sufficient to produce these revolutions and catastrophes, the traces of which are exhibited by its crust." We have seen how pole shift by itself can explain a wide range of geological phenomena such as ice ages, frozen mammoths, ancient myths, glacial deposits in presently tropical areas, mountain building, volcanic activity, magnetic-pole reversals, and animal extinctions. Not only have we seen that these various phenomena themselves are associated, we have also pointed to a *single cause* for them. We must also remember that psychics and scientists alike call for the occurrence of great earthquakes, increased volcanic activity, and sudden climatic change in the near future. Since some of these phenomena are already being observed, one must wonder what is now going on on our planet, especially since a major magnetic reversal is long overdue and some scientists interpret current magnetic-field intensity reductions as a prelude to such a reversal.

Bella Karish offered a new, intriguing hypothesis relating to this array of geological phenomena. In her model of the earth, she described just how all these phenomena related to one another. Karish said that many of man's present activities are causing geological disturbances that result in earthquakes, and for this reason the number of major earthquakes would steadily increase. In particular she singled out the detonation of nuclear bombs and changes in underground water levels. She said that these disturbances and earthquakes would increase the earth's wobble and this wobble in turn would cause minor land shifts and energize continental drift. These land movements would in turn bring on more earthquakes and greater wobble. She said things would go back and forth like this until the earth tumbled or flipped on its axis about 90 degrees. Thus she links man's activities with earthquakes, and earthquakes with wobble and plate movements, all of which in turn eventually lead to pole shift.

At the time Karish told me this, I hadn't read the many recent geological reports that support various aspects of this model. The crucial element was her appreciation of the fact that large earthquakes can actually jiggle the earth's axis of rotation and increase the wobble. So far as I know, Karish is the first psychic to bring together in coherent fashion the range of various phenomena that different psychics foresee for the future. In her model, the pole shift predicted for around 2000 A.D. is viewed as a natural consequence of accelerating geological activity brought on in part by our own actions.

If a polar shift were to occur, the movement would still be too slow to knock a person off his feet or send everything on the earth's surface suddenly tumbling. More likely, we could expect gradually worsening conditions and changes, where fear of what each hour will bring would be just as bad as the changes themselves. Imagine the following scenario:

> It is 5 P.M. in Boston and the people rushing home from work hardly notice how the setting sun seems to hang on the horizon. After several hours people start to wonder why darkness is not coming and they start to fear the faint dull roar they hear. Some also begin to feel light on their feet whether from the giddiness induced by the prolonged twilight or from subtle gravitational and magnetic changes that the earth's shifting in space is creating. Some have that I-could-jump-over-anything feeling. Animals act skittish and then suddenly all start to move or migrate in the same direction. Then the sky reddens as huge clouds of dust begin to blot out the sun. Next, a steady wind starts to blow. As the wind strengthens, the faint dull roar heard earlier grows even louder as if the source were moving closer. But just then, a temporary stillness sets in and the air seems like it is being sucked up by a giant vacuum cleaner. There is no sign of movement; all the animals are gone. After a few minutes, the winds are back even stronger. There are gusts up to 100 miles an hour. Trees are plucked out of the ground, and railroad trains are tumbled over and over and shuttled along like hockey pucks. As the wind jets up to over 200 miles per hour, buildings and everything above ground are decimated. The air becomes a thick mixture of dirt and debris. Those fortunate enough to be tucked safely away below the ground find the air hard to breathe since it is being drawn by the holocaust above. The wind-chill

factor has plummeted the temperature down to just above freezing, even though it is spring. There are almost continuous, grandiose electrical storms. Quakes and volcanoes are set off around the world and a rift opens up as the earth splits in several places to relieve the stress produced by the shift. This holocaust goes on and on as if it is never going to stop: ten hours, fifteen, twenty, forty, forty-eight. Then suddenly the winds subside and material from the sky starts to come crashing down. For a few minutes it seems to be raining automobiles, boats, washing machines, and kitchen sinks. The temperature comes back to normal for this time of year, a pleasant 50° F. But this temperature rise continues. By the next day, the third since the start of the shift, the temperature has hit 103° F, just what one would expect in equatorial Africa for this time of year—not Boston. From Boston to St. Paul to Seattle to Anchorage, out come the Bermuda shorts. Layers of mud spread for thousands of miles as a grim reminder of the holocaust. The decaying bodies of animals of every size and shape are found in caves where they huddled together in their last moments.

Karish's emphasis on the importance of nuclear detonation in pole shift and the turning up of the earth's internal burners should give us pause. Continued nuclear testing, the advent of the neutron bomb, the proliferation of nuclear power plants, proposals for the use of nuclear devices in construction and mining activities, and the ever-present threat of nuclear war quickly come to mind. For example, eastern metropolitan cities should note that a 1961 earthquake blew up a Russian nuclear power plant in the Ural Mountains that leveled an entire town and surrounding countryside. There is even a way that nuclear detonation could directly lead to the more serious consequences of a pole shift. Recall that some Princeton geologists took up the possibility that a rapid surge of one of the polar ice sheets would produce rapid cooling and bring on an ice age which, as has been discussed, could bring on pole shift due to an ice accumulation unbalancing the planet. Now consider what Walter Sullivan in *Continents in Motion* tells us: "One bizarre by-product of this surge hypothesis was a fear that some reckless government with nuclear energy at its disposal might set such a surge in motion." Sullivan reports that this potential danger was discussed at high levels of government. Geophysicist Dr. Gordon J. F. McDonald, who at the time was vice-president of the Institute for Defense Analysis in Washington and a member of the President's Science Advisory Committee, expressed worry that through nuclear explosions detonated along the base of an ice sheet a land-locked equatorial country could bring catastrophic change to the industrialized temperature-zone countries.

Before I began writing this book, I felt that while psychics had a lot of geology supporting their predictions of ever-increasing earthquakes, they were out in left field concerning pole shift. However, as I turned up more and more phenomena that could lead to such a catastrophe, I began to wonder. While I still can't point with certainty to the specific mechanism or combination of mechanisms that would cause a capsizing, I have become impressed with the likelihood of such a catastrophic event and its potential role as a key factor in geologic change, from mountain building and volcanism to continental drift and climatic upheavals. My judgment at the moment is that the case for eventual axis shift is at least as good as the case for increased earthquake activity in the near future. Whether it might occur by the end of this century is another question, but again I am inclined to accept the psychic view, if only to give impetus to further research.

If certain recent archaeological findings are correct and modern man has walked the earth for almost one million years instead of only thirty-five thousand years, then it seems that our present civilization sequence with its roots in the Near East a mere eight thousand years ago has been lucky to escape the type of monumental catastrophe that could well have erased all traces of many earlier civilizations. Philo of Alexandria gave us cause to wonder when he wrote, "By reason of the constant and repeated destruction by water and fire, the later generation did not receive from the former the memory of the order and sequence of events."

Perhaps catastrophic pole shifts will go on controlling the birth and death of civilizations until one civilization learns how to keep the earth from undergoing what may be a natural planetary process.

CHAPTER 8

Ruled by the Stars? The Astronomical View

"... we cannot rule out absolutely that there may be some reliable correlation between them [the planets] which could be used to predict earthquakes."

Dr. Roger N. Hunter,
Geophysicist, National
Oceanic Survey in May–June,
1971, *Earthquake Information Bulletin*

"... eventually some slight nudge will suffice to overcome the last scrap of the frictional hold [between tectonic plates] and bring catastrophe in one fast, tearing, scraping slide. What produces the final nudge though? Where does it come from? And when?

"In this book, Drs. Gribbin and Plagemann are on the trail of that nudge ... And on this subject they came up with the strange influence of the position of the planets in an odd (but rational) echo of astrological thinking."

Dr. Isaac Asimov in his foreword
to *The Jupiter Effect*, 1974

Evidence is mounting that there are periods when the earth is more likely than usual to experience earthquakes. Those periods seem to correlate with certain alignments and conjunctions of planets. There may even be rare times when the earth is susceptible to sudden vast geological change (as in the rise and fall of continents) and pole shift.

Some people believe that the planets and stars alone are the cause of such events. "When the moon is in the seventh house and Jupiter aligns with Mars"—that sort of astrological "New Age" thinking. Preposterous?

Well, maybe not. During the past few years, scientists writing in well-known and reputable journals have brought attention to the relationship between the movement of the planets and physical activity on the earth. The range of this activity is vast—from ocean and land tides to weather changes and the occurrence of thunderstorms, from changes in the earth's spin and magnetic field to interference with radio transmission. Not the least part of this activity is the triggering of earthquakes.

According to Dr. Livio Stecchini, who has taught this history of measurement at several leading American universities, astrology in its original form seems to have been based on theories that appear quite reasonable to modern science. Astrology can trace its roots back at least to the Sumerian and Babylonian civilizations, which date from the third millennium B.C.

The Sumerians and Babylonians paid a great deal of attention to the positions of the planets. Pliny, the famed Roman naturalist, wrote in his *Natural History* that "the theory of the Babylonians deems that even earthquakes and fissures in the ground are caused by the force of the stars, that is, the cause of all other phenomena, but only by that of those three stars [planets] to which they assign thunderbolts." The Sumerians and the Babylonians wrote prayers and hymns on their cuneiform tablets to these key planets. About Mars, which they called Nergal, the fire star, they wrote: "Radiant abode, that beams over the land, who is thy equal?" "The heavens he makes dark, he moves the Earth off its hinges." "Nergal . . . on high stills the heavens . . . causes the earth to shudder." Babylonian astrologer-priests carefully noted the movements of Mars, Venus, Jupiter, and Mercury in their temples and stepped pyramids. In fact, scholars wonder how the Babylonians made many of their astronomical observations of the planets and their satellites without the aid of telescopes. (Were the Babylonians only carrying on activities passed to them by some far older and far more advanced civilization of which we presently have no trace—a civilization wiped out by a pole shift?)

Whatever their origins, the concepts of astrology have persisted through the ages, and recently scientists with impressive credentials have turned their attention to the relationship between the planets and earthquakes. Dr. R. Tomaschek of Munich, writing in the geophysics section of *Nature* (1959), states that "... a remarkable correlation between the position of Uranus and the moment of great earthquakes has been established ..." Most recently Drs. John Gribbin and Stephen Plagemann took up this issue. Gribbin is an editor of *Nature* and holds a doctorate in astrophysics from Cambridge University. Plagemann holds a doctorate in physics, also from Cambridge, and is a researcher with NASA. While at Cambridge Plagemann worked at the Institute of Theoretical Astronomy under the eminent astronomer Sir Fred Hoyle. In their book *The Jupiter Effect*, published in 1974, Gribbin and Plagemann say, "Now, to the surprise of many scientists, there has come evidence that ... the astrologers were not so wrong after all; it seems that the alignments of the planets can, for sound scientific reasons, affect the behavior of the earth."

Gribbin and Plagemann believe that rare planetary alignments can set off periods of extensive earthquake activity on a worldwide basis. They believe that the extra "pull" of the planets can trigger incipient earthquakes in areas where stresses have been building up for a long time. The relatively small additional pull of the planets acts like the straw that breaks the camel's back. They predict that the coming rare alignment of planets in 1982 "will trigger off regions of [unprecedented] earthquake activity in the earth." They say that in 1982 the planets will be in "superconjunction." Furthermore, "Between 1977 and 1982 the planets will be moving into an unusual alignment in which every planet will be in conjunction with every other planet; all the planets will be aligned on the same side of the sun."

The term "alignment" is used loosely by Gribbin and Plagemann since the planets will actually span an arc of some 60 degrees on the same side of the sun. Nevertheless, they predict that this positioning, which occurs only once every 179 years, will probably trigger earthquakes along California's San Andreas Fault. They note that preceding the last time such an alignment occurred, in 1803, California underwent a period of concentrated earthquake activity. They quote a leaflet from Mission San Juan Bautista near San Francisco, the oldest California mission to have been in continual use since it was founded, in 1797: "In 1800, between October 11th and October 31st there were as many as six shakes per day." Though records from the early period are far from

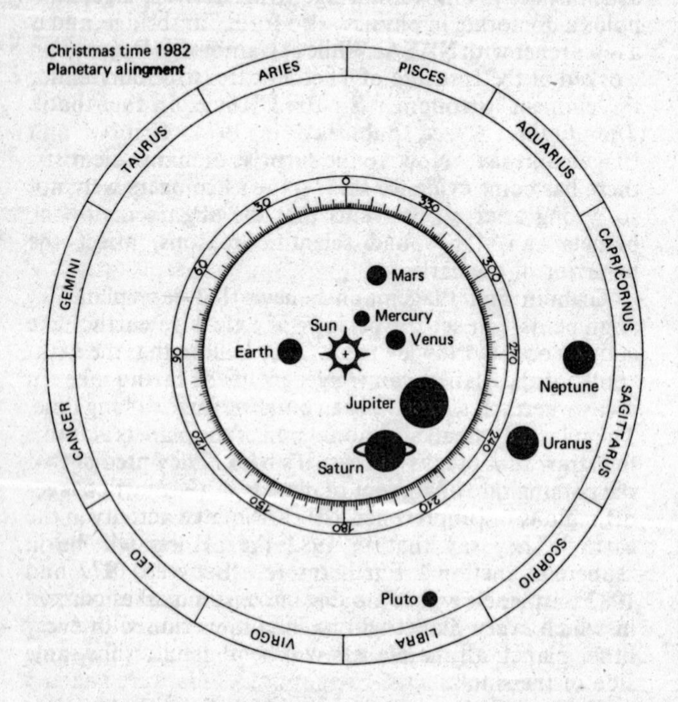

The planetary alignment in late 1982, at about Christmas, indicates a pattern that could cause great sunspot activity and enormous earthquakes.

complete, Gribbin and Plagemann note that these earthquakes are listed in the official *Earthquake History of the United States*, published in 1973 by the National Oceanic and Atmospheric Administration (NOAA) Environmental Data Service. In addition to the earthquakes at San Juan Bautista the NOAA publication lists another major shock that occurred about this time in the San Diego area (November 22, 1800). It damanged the walls of adobe buildings in San Diego and cracked walls of a new church under construction at San Juan Capistrano. A violent quake was also felt in 1800 at Santa Barbara. "Thus," say Gribbin and Plagemann, "we have a record of at least three intermediate (greater than 6) size earthquakes near the grand alignment of 1803." They go on to note that other than the 1906 San Francisco quake, there has been little major quake activity in northern California during the last 200 years. They also admit that the year 1800

> does not quite fit the date of the last planetary alignment (1803), but as we have already mentioned, it is not clear whether it is the exact alignment or some equally rare configuration of the planets in the years building up to the alignment that provides the trigger; and if the fault is ready to go, the steadily increasing level of activity caused by the alignment effect could set it off even before the maximum trigger effect was produced.

Gribbin and Plagemann discuss how they were led to their conclusions. The key factors are: (1) A study of the effect of the planets on land tides, where the ground moves by as much as four inches up and down over a twelve-hour period, much like ocean tides. (2) Recent NASA studies that show that most moon quakes occur when the strain associated with solid tides caused by the earth's gravity is greatest. (3) A Russian study showing that a series of small earthquakes in Nevada were linked to the tidal influence of the moon. (4) A study by Dr. T. Simkin of the American Museum of Natural History in

1968 that showed that earthquakes associated with volcanic activity and the collapse of a volcanic caldera occurred *only* at each and every extreme of the local ocean tide. (5) Studies that show that solar flares affect the earth's spin. (6) Studies that show that the sun's activity can cause unusually bad weather. (7) Correlations between the solar sunspot cycle and the alignment of the planets. For example, when another planet (Venus, Earth, or Jupiter) is on the same side of the sun as Mercury during Mercury's closest approach to the sun, Mercury's effect on solar activity is twice as great as when this planet is on the opposite side of the sun. (8) A detailed study on sunspot activity made in 1972 by Professor K. D. Wood of the University of Colorado that showed that sunspots follow a repeated pattern roughly every one hundred seventy to one hundred eighty years. Dr. Wood and other scientists have predicted that a sunspot maximum will occur early in 1982. (9) A study by Dr. John Nelson of RCA Communications that found planetary alignments to be significant predictors of radio "weather." For example, when any three of the nine planets are aligned, radio disturbances occur even when there are few sunspots. When five or six of the nine planets are in alignment, more severe radio disturbances occur. The greatest radio disturbances occur with more complex geometric alignments. Follow-up studies on Nelson's work showed a 180-year period of disturbances due to resonances between the planets, in which Jupiter, the largest of the planets, plays a key role.

If Gribbin and Plagemann are right in their basic premise, then the planetary positions for the year 1982 should cause us less concern than those of 2000, the year the psychics predict pole shift will occur. Reporter Tom Valentine, in his book *The Life & Death of Planet Earth*, tells how, with the help of astronomer Dr. Harvey J. Augensen of Northwestern University, he calculated the positions of the nine planets relative to the sun for May, 2000. Valentine says that in 2000 a much more important alignment will occur. At that time Earth will be all by itself on one side of the sun. Heading directly away from

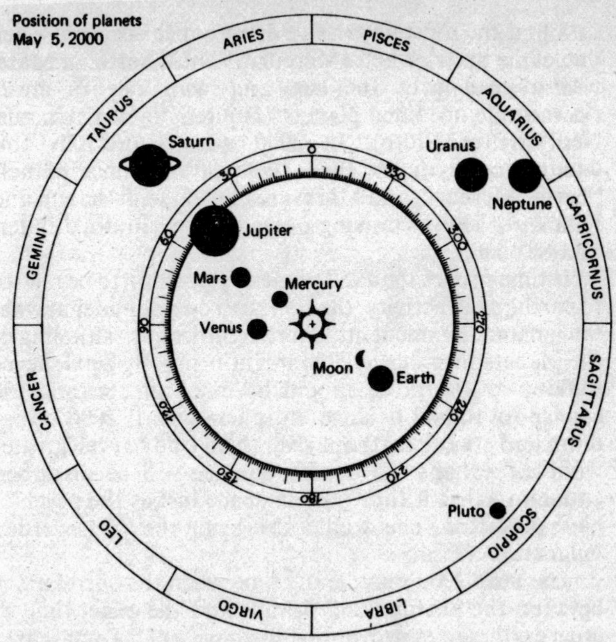

The alignment of the planets for May 5, 2000, indicating a pattern that could cause great planetary disturbances and enormous earthquakes.

the Earth in a straight line on the other side of the sun will be Mercury, Mars, Jupiter, and Saturn. Even the moon and Pluto have a place in this alignment. According to Nelson's radio weather studies, because more planets are in alignment greater disturbances should occur than in 1982, especially since the solar system's two largest planets—Jupiter and Saturn—will be in alignment. Gribbin and Plagemann say that an unpublished NASA study shows that increases of up to 20 percent in solar activity occur when just Jupiter and Saturn are in alignment. Furthermore, the 2000 alignment duplicates the more complex geometric condition which Nelson found to bring the greatest radio disturbances. Nelson said that the most impressive disturbance occurred when one of the inner planets (Mercury, Venus, Earth, or Mars) was aligned with the sun and with one or more slower-moving outer planets (Jupiter, Saturn, Uranus, Neptune, or Pluto). In 2000 we get precisely this condition. No fewer than *three* of the inner planets (Earth, Mercury, and Mars) are aligned with the sun and with *three* slower moving outer planets (Pluto, Jupiter, and Saturn).

If time proves the 1982 planetary position to be related to earthquake activity, then one can only shudder at what the greater alignment and more significant positioning of the planets to occur in 2000 might bring. By 2000 the ice buildup at the poles should be much greater and the earth's rotational balance more tenuous. If 2000 brings more and greater earthquakes, which lead to even greater wobble, perhaps the earth's balance will be disturbed enough so that it flips over in space just as the psychics have predicted. This would also bring the long-overdue magnetic reversal.

The year 2000 may be the time when the correlations between the positions of Uranus and the exact time of great earthquakes prove "highly significant." Tomaschek, whom I quoted in part earlier, says

> a remarkable correlation between the positions of Uranus and the moment of great earthquakes has been established for a certain period (1904–06). Gutenberg and Richter's data of *all* earthquakes equal or greater than magnitude of 7¾ have been used . . . A total of 134 earthquakes have been investigated. In this a fairly significant amount of cases have been found, where Uranus was very near its upper or lower transit of the meridian of the epicenter in the time of great earthquakes.

Uranus is relatively stationary with respect to the earth, and for purposes of illustration we can consider Uranus as a fixed point in space. Then, since the earth revolves on its axis, Uranus is directly above or below each place on the earth twice a day. In other words, Tomaschek has found that the location of an earthquake is within 15 degrees east or west of the line drawn from Uranus to the earth at the time of an earthquake. The probability of Uranus being in this position is only one out of six, and thus if the times of occurrence of earthquakes are distributed by chance, a uniform distribution of the positions of Uranus can be expected. While Tomaschek found a chance distribution for the positions of the sun and moon at the time of the earthquakes, he found the positions of Uranus to be quite otherwise. For the most part the Uranus earthquake correlation was just better than chance over the forty-nine-year period studied, but for the 1904–06 period fifteen out of the twenty-three cases (65 percent) showed a unique alignment for the positions of Uranus at the time of a very large quake. Chance would give only 3.8 cases. The probability for this occurring by chance is 0.000000269, or approximately one in a hundred million. Says Tomaschek, "The years 1904–1906 correspond to a conspicuous maximum of energy released by the earth through earthquakes. The earthquakes in the period were of shallow, intermediate and deep origins."

It is interesting to note that in the eight out of twenty-three cases where the Uranus relationship didn't occur, the quakes were all of shallow origin. Thus Tomaschek writes, "To state the plain facts: one is led to the conclusion that the position of Uranus within 15° of the meridian at the moment of great earthquakes can be regarded as significant and that there exist times of longer period (several years) when it is very highly significant." Tomaschek goes on to note that after 1906 the correlation drops "but it remains greater than average," and as an example he shows how it held for three of the most remarkable earthquakes in this century: the 1923 Tokyo, 1933 Honshu, and the 1950 Assam (India) quakes, "when the position of Uranus had already greatly changed," from that of the 1904–06 period.

Perhaps this special and occasional influence from Uranus will be the proverbial straw that breaks the camel's back, the straw which will cause the earth to tumble during 2000. That year is already expected to bring more and greater quakes, and earth wobble is expected to increase as a result of other planetary positions. That is also the year in which the earth's rotational balance is likely to be compromised by polar ice buildup. *It can now be seen that many different factors and forces seem to be coming together in the year 2000. Their cumulative effect could bring on a pole shift.*

Cayce and my psychic group all clearly noted the influence of the planets on earthquake activity. Abrahamsen guided by our discussion and my questions was able to give information on the exact geometric positions and activity of the planets at the times of major earthquakes. At first this information sounded like it was merely an embellishment of Nelson's work at RCA concerning planetary alignments and radio weather. But as I noted in Chapter 1, I became more of a believer after the occasion in which Abrahamsen said that John Wilkins, an obscure seventeen-century astronomer, "came up" to help him, when the history books confirmed the data about Wilkins.

Abrahamsen also stood up to an even tougher test when we made the December, 1972, readings. In a reading Abrahamsen first goes into a meditative state before he responds to questions. Before I began to ask him specific questions about planetary relationships, I asked him a control question—a technique that I developed and used in my archaeological work with him. If, at the beginning of a reading, Abrahamsen couldn't answer a specific question to which I already knew the answer, there was no sense in proceeding to ask for new and detailed information on other aspects of the same subject. My control question for Abrahamsen at the beginning of the planetary reading session was: "Will you give the relationship between Mars and the earth in relationship to the sun on April 18, 1906, the day of the great San Francisco earthquake?" Abrahamsen answered quickly and accurately. The celestial ephemeris showed that the relationship he gave between the heavenly bodies on this date was accurate: Mars was about 1.52 astronomical units distant from the sun as compared to the earth's 1.00 astronomical unit, and the angle between Mars, the sun, and the earth was approximately 42 degrees. Since Abrahamsen had answered the control question correctly, the reading proceeded and I had some confidence in his answers to my questions.

In this reading Abrahamsen proceeded to give complex geometrical relationships between the planets at the time of very large earthquakes. A series of triangulations was indicated between Mars, Uranus, the earth, and the sun. And then Abrahamsen went a step beyond. He said these triangulations only set up the possibility for great quakes. A quake occurred only if there also was a temporary deviation of Mars's orbit in which Mars moved fifty to one hundred thousand miles closer to Earth. In other words, this deviation of Mars acted as the trigger. If it wasn't pulled, then all was safe and there was no quake, despite the positioning of the planets. When this deviation takes place, Abrahamsen said, Mars's orbital speed also temporarily increases and Mars's axis tilts. Abrahamsen indicated that before this temporary deviation Mars gives a warning by temporarily increasing its wobble and temporarily increasing its reddish appearance (beyond the usual seasonal changes). Abrahamsen's statement brings to mind the Babylonians, who believed that Mars contributed to the cause of earthquakes and who called Mars "the unpredictable planet." It also brings to mind the strange fact that they noted Mars's rise, setting, disappearance, return, its position in relation to the equator, changes in its illuminating power, and its relationship to the other planets.

Abrahamsen spoke about Uranus, too. He said that Uranus also underwent a temporary change at the time of a quake. Uranus temporarily increased its orbital velocity, and the greater the temporary acceleration of Uranus, the larger the earthquake. In a later reading he indicated that if the positions and declinations of Venus and Jupiter were added to the original triangulations he gave, the timing of even more quakes could be correlated with planetary positions. He said this would correlate with moderate-sized earthquakes as well as the largest quakes. Then, for even greater refinement in calculation, he said that the positioning of the moon and the positioning of Ariel (one of Uranus' satellites) could also be used.

In short, things got very complex regarding planetary activity and quakes. So it was with a sigh of relief that I received a phone call one day in 1973 from Dr. William Kautz of the Stanford Research Institute (SRI), whom I mentioned earlier. Kautz asked if he could run the planetary-earthquake data I had collected on the SRI computer. His doctorate is in mathematics and he has more than twenty-five years' experience in computer work. He himself had been doing research on possible relationships between earthquake activity and planetary position, and when he learned from associates of his at SRI of my work with Abrahamsen he thought it would be worthwhile and easy enough to check. Kautz spent several days with me at my home in Tucson, going over what I had. But soon after Kautz returned to SRI he realized that checking Abrahamsen's information out wasn't going to be so easy after all. So he asked if he could work directly with Abrahamsen to get more information regarding the mathematically complex relationships Abrahamsen had given.

Kautz has spent a good deal of time by now working with Abrahamsen, but as of November, 1977, he is not ready to publish his findings. One reason is that no observations have been made on short-term changes or sudden deviations in Mars and Uranus. But Abrahansen's information offers open-minded scientists new directions for data collection and analysis. Abrahamsen, in his psychic state, might have accurately seen the actual workings of the planets at the time of earthquakes.

Despite the lack of data in this aspect of the planets-earthquake relation, we can get a sense of the accuracy of Abrahamsen's statements. In the readings we did together, he presented a way of viewing the universe and our solar system quite different from professional astronomers, who believe that the laws of gravity are the most important way to understand the movements of the planets. Abrahamsen said that the laws of magnetism are even more important. He pictured the solar system as a giant electromagnetic field, in which lines of magnetic force stretch out between the sun and the planets. He said that the planetary system is carefully balanced and each planet acts like a magnetic gyrocompass, where the sudden deviation of one planet almost immediately and very directly affects all the other planets through this magnetic field. Thus, according to Abrahamsen, small planets, which have small gravitational effects, could nevertheless have large (magnetically derived) effects on the other planets when they were located at pivotal positions.

Under this system we can see how a temporary deviation in Mars could cause a disturbance on earth. That Mars was singled out is interesting when we consider that Mars, and only Mars, bears a number of striking similarities to the earth in regard to its balance in space. The axes of rotation of both Mars and the earth are inclined, or tilt, at approximately 24 degrees. The earth's angle of tilt varies from 22 to 24.5 degrees over a 40,000-year cycle; Mars's tilt varies likewise over a 50,000-year cycle. Not only do both planets have approximately the same axis tilt and period of tilt, they also have approximately the same mean time of axis rotation—that is, length of day. The average duration of the earth's day is 23 hours, 56 minutes, and 4 seconds; Mars's day is 24 hours, 37 minutes, and 23 seconds. With these similarities, and operating in an electromagnetic field, it is not hard to imagine that a sudden increase in Mars's wobble would cause the earth to undergo a similar increase in wobble and possibly an earthquake when Mars was at a pivotal position in the electromagnetic field of the solar system. The increase in the earth's wobble that Mansinha and Smylie found accompanying large earthquakes, discussed in the previous chapter, comes to mind, especially since Mansinha and Smylie found that the earth's pole path and wobble changed five to ten days *before* most of the large quakes. It seems as if Mars's deviation and wobble is resonantly picked up by the earth through the self-balancing quality off an electromagnetic field, like one magnet moving another. This change in the earth's equilibrium in turn sets off an earthquake. Such an earthquake, if other critical factors were present in maximum force, could provide enough of a jolt to disturb planetary wobble and precipitate a pole shift.

Also of interest is the fact that Mars, far from being geologically dead like the moon, as was once thought, is internally alive and active much like the earth. There are a number of geological forms similar to our planet's, such as volcanoes, canyons, giant dusty basins, jumbled uplifts and fractures, and dry arroyos with intricate tributaries. The Mariner sensors have even detected evidence of great cold fronts similar to our atmosphere's and surface temperatures that range from –190° F. at Mars's frozen polar caps to a balmy 80° F. in the equatorial region. And in late 1976 the Viking lander on Mars showed it to be a lively place indeed, as its seismograph detected two quakes of at least 6 on the Richter scale.

In addition to Abrahamsen picking Mars as a key planet in relating planetary positions to earthquakes, his identification of Uranus' role is also of merit when we consider Tomaschek's discovery of the correlation between the position of Uranus and the time of earthquakes. In support of Abrahamsen's concept of the solar system, Tomaschek (who I presume knows nothing of Abrahamsen) feels that this correlation between Uranus and quakes only occurs for certain periods of time because of "the fact that Uranus is the only planet of which the direction of its axis of rotation coincides with the plane of its orbital revolution. A possible magnetic field would influence the solar plasma in a way quite different from all the other planets." The term "solar plasma" refers to the electromagnetic field emanating from the sun.

Abrahamsen is by no means out on a limb when he speaks of the electromagnetic nature of the solar system.

In addition to the support he gets from Tomaschek, the views and findings of a number of other researchers back him up to some degree. Gribbin and Plagemann point out that interplanetary space is filled by a plasma, a gas in which the atoms have been stripped of electrons, leaving charged atoms (ions). This plasma moves outward from the sun, its place of origin, and is a very strong conductor of solar electricity and magnetism. Solar flares emanate from regions in the sun with unusually strong magnetic fields, and these flares cause the earth's magnetic field, which extends far into space, to fluctuate. At times of solar-flare activity, magnetic storms range all around the world. Besides solar flares and the earth's magnetic field, the electromagnetic nature of the solar system and indeed the entire galaxy is indicated by arched solar coronal structures, by repulsion of comet tails by the sun, by the polar lights, by the high charges of cosmic rays, and by the magnetic properties of certain stars and planets. Still other indicators are polarized light from certain stars; the electromagnetic signals coming from space; radio noises from Jupiter, Venus, and Saturn; quasars and pulsars; the relationship of disturbed radio transmissions to planetary configurations; and the fact the earth's volcanic activity is greatest along the equator, where its magnetic field is the weakest.

Of particular interest here is Jupiter's magnetic field. Scientists have long known of it, but in 1974 the Pioneer 10 unmanned spacecraft, which passed by Jupiter, shocked astrophysicists when it showed Jupiter's magnetic field to be far greater than expected—so strong, in fact, that its penetrating electrons almost destroyed the craft. Scientists were also amazed when Pioneer 10 showed that Jupiter's magnetic tail (a bulge in its field caused by the solar wind) was even longer than the earth's long magnetic tail. Jupiter's tail was found to extend 430 million miles; by comparison, Earth's tail is 395 million miles. Jupiter's tail extended the distance between it and the orbit of adjoining Saturn. And since Abrahamsen's reading (December, 1972), the Mariner 10 spaceprobe in 1974 surprised scientists by showing that Mercury also had a magnetic field, the acquisition of which is difficult to explain.

Most baffling to scientists is the fact—shown by Pioneer 10—that Jupiter's magnetic field can suddenly vanish. Dr. Robert Nunamaker, the Pioneer mission commentator, said, "All previous theories of what might be happening out there are being changed moment by moment." Perhaps the reason behind Jupiter's vanishing magnetic field can explain why Uranus seems to be associated with earthquakes only for certain periods of time.

Not only are scientists discovering more about the electromagnetic nature of our solar system, but new, startling data require them to revise other notions about the nature of our solar system. For example, we have known only since 1976 that Venus and Mars are still geologically active, and that Mars has a volcano bigger than the entire state of New Mexico. We also learned even more recently that there are rings around Uranus rivaling those of Saturn.

In view of these and other major new facts just now being gathered about the solar system, I urge an open-minded attitude in those astronomers who are quick to say there is no possible relationship between planetary activities and earthquakes. The rare planetary positions due in 1982 and 2000—when considered in terms of possible damage to civilization and lives lost—should warrant the most serious scrutiny. Anything less than that would be criminally irresponsible, and perhaps even suicidal.

CHAPTER 9

Biorelativity: You Are the Quake

"Because of the evil deeds of the land, the ground is parched, no rain has fallen upon it..."

Jeremiah 14:4 (Lamsa Bible)

"...there are those conditions that in the activity of individuals, in line of thought and endeavor, oft keep many a city and many a land intact, through their association of the spiritual laws in their associations with individuals..."

Edgar Cayce
(311-10, November 19, 1932)

"We've become a race of technologically advanced imbeciles living in a world we don't understand and don't have any real desire to know anything more about than what affects us directly and individually."

Herbert Allen Boardman
in *The HAB Theory*, by
Allen W. Eckert

My work with psychics has been mind-boggling. I have learned many extraordinary facts of esoteric science and have had many insights into the nature of reality. But the most extraordinary thing of all I learned is that human behavior, social activity, and human thoughts play an important role in the occurrence of earthquakes and climatic changes.

This brings to mind a number of biblical themes, such as reaping what we sow. Cayce, Abrahamsen, Karish, and Elkins often referred to "thought forms." They said that the various activities and attitudes of man create thought forms of different sizes, shapes, and colors. These thought forms are said to be physically definable as "vibrations."

All matter vibrates, from the densest metal to the most rarefied gas. Vibration is also the common denominator of light, sound, magnetic, and electrical energies. In the relativistic world of Albert Einstein, matter can be converted into energy and vice versa.

Psychics describe thought forms as a mental substance, something intermediate between matter and consciousness. Our thoughts are "formed" from psychic energy, a kind of energy that is presently unrecognized by science but that is actually a fifth force in nature.* From the psychic point of view, our thoughts extend beyond our bodies and into physical space. And since energy can be neither created nor destroyed, the energy of thought, psychics say, still exists as a sort of atmosphere or field surrounding the planet, recording all the experience of humanity. This is the so-called akashic record, which Cayce and other psychics claim to "read" when they obtain paranormal information about the past. While the reality of such a "field of mind" is not yet demonstrated empirically, it has been proposed (in the 1973 *Science Yearbook*) as a conceptual tool by psychologist Dr. Elmer E. Green of the Menninger Clinic in Topeka. And the late eminent neurophysiologist Dr. Wilder Penfield also approached this notion in his final statement, *The Mystery of the Mind*, published in 1975. Wilder's decades of research into the relation between brain and mind led him to conclude that even the highest brain mechanisms cannot explain the nature of the mind: ". . . the mind is peculiar." "It has energy. The form of that energy is different from that of neuronal potentials that travel the axone pathways."

* For a useful survey of the evidence suggesting the reality of psychic energy and how it functions to perform paranormal phenomena, see *Future Science*, edited by John White and Stanley Krippner (Garden City, N.Y.: Doubleday Anchor, 1977).

We can see that serious consideration is being given to the possibility of a new force in nature that is directly influenced by man. I call this *biorelativity*, the interaction of people with their physical environment via psychic or mind energy.

This sounds foolish to skeptics, of course, but it is nevertheless a fact that people give off various energies. That is why polygraphs (lie detectors) work. But the energies of the human body don't stop at the skin. The London University biologist Dr. Lyall Watson, in *Supernature*, discusses the work of several researchers who have used sensitive voltmeters to measure changes in the electrical field given off by the body. These changes can be measured up to several meters away.

Recently, via the process of Kirlian photography, this field has even been seen. Kirlian photography is a lensless technique developed by Russian scientists and now widely used around the world. In it, the radiation of the human body is used to expose the field. A luminescence of changing colors and shape is seen around the body, forming an aura or corona of exquisite beauty. This luminescent field has been found to vary with a person's emotions and state of health. The Kirlian field has been likened by some to the human aura pictured by the ancient Egyptians, Greeks, and Indians, which some psychics say they can see. Also, some feel there is a correlation between the many points of emanation of this Kirlian field around the body and the Chinese acupuncture points now being so actively studied and used by Western scientists.

The most convincing evidence for such an electromagnetic field, admittedly a not very intense one, comes from the direct magnetic measurement of this field by researchers at the Massachusetts Institute of Technology, using a highly sensitive magnetometer in a specially shielded room. "The field around the heart has been found to be about 5×10^{-7} gauss. This was one millionth of the earth's steady magnetic field and one thousandth of the fluctuating background of the earth plus typical city noises in a bandwidth of 0 to 40 hertz," they report. The field around the abdomen was occasionally noted to be higher than 10^{-6} gauss.

This magnetic field about the human body has been found to vary from person to person and to vary with changes in the mental state and behavior of the person. The Russians have made similar measurements. The MIT researchers verified the authenticity of these magnetic changes by the comparison of their magnetic measurements with electrocardiograms (EKG) and electroencephalograms (EEG), where parallel patterns were produced. For example, the magnetic field above the head was found to vary in correspondence with the alpha rhythm of the subject's brain. Hence we see that there is indeed some physical basis for the concept of thought forms.

From this point on in my discussion, I will present thought forms as if they were electromagnetic vibrations. However, I should point out that this marks a divergence on my part from the mainstream of psychic tradition, which considers them to be other than—and more basic than—electromagnetism. I also must point out that we have no idea what environment-influencing information, if any, may be encoded in these electromagnetic vibrations. We can measure the magnetic field of the body or the electrical rhythms of the brain, but these say nothing directly about the meaning of the message they might be carrying. It's like measuring the electrical energy traveling along a wire. If Morse code is being transmitted, we might pick up the length of the dots and dashes, but unless we understand the code we won't understand the information being transmitted.

The electromagnetic characteristics of a person's emanations must resemble the electromagnetic characteristics of the solar system which Abrahamsen and others consider so important. Human electromagnetism fields could travel out and interact with similar fields in the environment much like tuning forks of the same wavelength responding or resonating to each other.

The psychics said that the thought forms given off and created by man interact with the factors behind earthquakes, volcanoes, and geological activities, as well as the factors behind climatic change. They also said thought forms could either increase or decrease rainfall, much as the American Indians believed in their rain dances. In the case of an earthquake-prone area, thought forms could cause temperature or pressure changes and thus affect the occurrence and nature of the earthquakes that hit the area. For example, the psychics said that if the people in an area lived "harmoniously," then the types of thought forms produced would interact to reduce the severity of earthquakes in the area. On the other hand, if the people in an area lived in "turmoil," then thought forms produced would tend to increase the severity of the earthquakes in the area. We thus get a situation where thought forms can modify the way in which the geological energy of an area is released—either positively through small, harmless tremors or negatively through large, damaging quakes.

The closest modern science has come to noticing the effects of thought forms has been in some of the recent controlled experiments on the effect of prayer on plants. In these, plants responded with increased growth to positive prayer, loving thoughts, or classical music, while they responded with decreased growth to negative prayer, hateful thoughts, or raucous music. Delawarr Laboratories in England and independent researchers in various countries in dozens of experiments found that seeds which were made the focus of positive prayer showed a 5 to 52 percent growth advantage (depending on the individual doing the praying) when planted. Dr. Bernard Grad, a biochemist at McGill University in Montreal, not only found that "treated" seedlings grew faster and the plants weighed more than those in a control group, but he also found that the plants from the "treated" seedlings showed greater resistance to insects. Impressed with Dr. Grad's technique and results, Dr. M. Justa Smith, a biophysicist at Rosary Hill College in Buffalo, New York, demonstrated that an enzyme "treated" by the same healer Dr. Grad used in his experiments showed significantly more activity than the "untreated."

Recently physicists may have also touched upon an aspect of thought forms in studies of individuals who can move, bend, break, and otherwise affect objects with their minds. This is called psychokinesis (PK). Israeli showman-psychic Uri Geller has drawn the public's attention to his seeming ability to bend objects at will. While his theatrics have confused this issue and the excellent research done on him by Stanford Research Institute,* most people don't realize that scientists have located hundreds of other people who can also bend or break objects at will. Many are children. Recently Dr. Lyall Watson of London University described to me a ten-year-old girl he studied who was able repeatedly to turn tennis balls inside out for him with her PK power alone. In Russia a number of internationally respected scientists have watched several female psychics move objects and deflect compass needles solely by thought.

More pertinent to the issue of man affecting the causes of earthquakes is the fact that a New York City artist and psychic named Ingo Swann, in experiments conducted at City College of New York, was able to raise the temperature of graphite blocks placed inside a Thermos container at the opposite end of the room. If this can happen, it is not far-fetched to think that the PK effect of a number of individuals acting in concert would be able to affect rock stress. It would be interesting to conduct experiments in which rocks are stressed in a hydraulic jack and then direct thought at them in order to get them to fracture and fail.

Through these examples I have tried to justify the concept of "biorelativity," that is, that thought forms might affect the material environment. The question we must now ask is: has this ever happened in a significant way? Based on research conducted as part of my graduate

*See Harold Puthoff and Russell Targ, *Mindreach* (New York: Delacorte, 1977).

studies, my answer is yes. In this research, I found that the large prehistoric populations in the American Southwest that I studied failed when their activities became "disharmonious" with the environment. I considered activities such as open warfare disharmonious. Over a period of seven hundred years as the activities of each of these populations became disharmonious, their rainfall and/or their water table dropped until they had to move away. On the other hand, the rainfall and the water table of nearby populations who still were in harmony with the environment (as indicated by the absence of evidence of warfare) did not drop, even though they were still in the same meteorological region. A number of diverse theories have been used to explain the climax of each of these large populations but my simple yardstick of "good guys–good environment/bad guys–bad environment" works for each and every climax population. This biorelativistic hypothesis even explained why one population in the region thrived throughout the time period studied.

As part of this research I made an effort to investigate one possible mechanism involved in the biorelativity process. I conducted two simple controlled experiments, one in the laboratory and one in the field. I can't remember who was more shocked by the positive results of these experiments—my professors or me. In both experiments the evaporation rate of water in shielded graduated cylinders increased when the water was subjected either to prayer or to the playing of "harmonious" music. This was not the case for the controls. Unfortunately, due to Arizona's dry weather, I was unable to measure differences in rainfall at the data sites, which were just 100 feet apart in the field experiment. A more complete account of these experiments is given in Appendix C.

While I was at the University of Arizona, I wrote two papers on this research and the concept of biorelativity. One paper was for a graduate course in Past Climate; the other was for a graduate course in Southwestern Archaeology. To my surprise both papers received A's and both professors felt that I had laid out a viable hypothesis worthy of further research. But when another professor heard about what I had written, he became wildly hostile and emotional. After he regained coherence he asked me why my theory involving rainfall and water tables didn't hold for the 1970s. I said that the drilling of deep wells eliminated a large population's dependency on yearly moisture conditions in the Southwest. I also said that I was sure there were many other factors involved and that what I did merely constituted a pilot study. Still upset, he stormed off.

During the week that followed this confrontation, something very interesting happened. It involved a peace march in Tucson. The march didn't begin until noon. All morning there hadn't been a cloud in the sky, but by mid-afternoon, after the march was over, clouds formed and there was a sprinkle. This was the first rain Tucson had received in more than 100 days. The next morning I saw the hostile professor again and asked him if he had taken note of the giant rain dance in the form of a peace march. Once again he stormed off. I wonder how many other possible human-environment interactions we fail to notice. And I wonder if a city can save itself from earthquakes like Nineveh did in Jonah's time simply by changing behavior and attitudes. If there really has been physical and social evolution, man may now be working on the evolution of an ethic—an ethic that may prove to be fundamental for the future of the race.

CHAPTER 10

Earthquake Prediction: Consult Your Local Goldfish, or How a Cockroach Could Save Your Life!

Someday a cockroach may save your life. Those clever little masters of survival have lived on earth for more than 250 million years. They are extremely adaptive and sensitive. They can stay in a very small area without food or water for half a year. Dr. Ruth Simon, a geologist with the USGS, recently had tiny refrigerated homes built for some. In December, 1976, three of these homes were placed along the San Andreas Fault in California. The homes have sensors on the floor that record every movement the cockroaches make. According to Simon, "Any disturbance in the cyclic rhythm that the cockroaches are accustomed to will disturb their behavior. Simon anticipates that the cockroaches will start behaving differently just before an earthquake.

While Ruth Simon watches her cockroaches, members of the Japanese Namezu No Kai (Catfish Club) dutifully watch their catfish. The group took its name from the legendary catfish that reside under the earth and create tremors when angry. The hundred-plus members of the club are spread across Japan and have successfully predicted several earthquakes by watching their fish swim about.

Is catfish-watching legitimate? Dr. A. J. Kalmijn, a marine biologist and physicist at the Scripps Institute of Oceanography in La Jolla, California, says, "I am convinced that the Japanese findings are valid." Kalmijn has been conducting his own research with catfish, testing their reactions to electrical currents introduced into their tank. He concludes that the fish can pick up minute changes that take place in the earth's electrical field before a quake. He stated that the changes agitate the catfish, causing them to react violently to noise and vibration. "I suppose that a person could keep a catfish in a tank at home to warn him of earthquakes. You periodically tap the tank and if the fish jumps about there is a good chance an earthquake is on the way."

This is what Japanese scientists found to happen with fish kept for this purpose in a laboratory in a quake zone. Kalmijn says that the ordinary catfish can sense earthquakes six to eight hours before they happen. Dr. Lyall Watson, noted earlier, reports that Japanese living along a highly fractured zone have also kept goldfish as an early warning system. "When the fish begin to swim about in a frantic way, the owners rush out of doors in time to escape being trapped by falling masonry." Watson believes the water conducts the low-frequency vibrations (7 to 14 cycles per second) that precede earthquakes.

On December 14, 1976, behavioral scientists at Stanford University's School of Medicine and Dr. Seymour Levine of the Stanford Primate Center announced that they had one dozen "psychic" chimpanzees. These chimpanzees, they said, amazed them by predicting two earthquakes. The scientists, who were conducting animal studies at an animal research center just two miles from the San Andreas Fault, recorded the significant behavior changes the chimps underwent a few days before two small quakes struck the area. Dr. Helene C. Kraemer, a medical statistician, said that "their behavior change was so significant that it seems unlikely it was due to chance... We believe we [have] found... scientific evidence that animals' behavior changes before earthquakes."

Science seems to be catching up with common sense. Many people know that animals can detect danger long before humans can, and that when they do so, they become restless. Many kinds of wildlife have been reported to act unusual just before a quake struck their range. Hours before the terrible 1975 Italian quake, the Washington Post News Service reported,

* * *

> Dogs barked uncontrollably... cattle in barns bellowed and tore at their chains... deer flocked down from the mountains, crowding together as if in terrible fright. Caged birds flew amok, hammering against their wire prisons. Cuckoos cried out in the fields, sounds never heard at night. Fowl refused to roost and made such a din that farmers thought their henhouses had been stormed by foxes. Rats and mice left their holes and ran free because all the cats had fled the village for the countryside.

Soviet scientists have gone so far as to establish animal warning centers in the quake-prone Uzbekistan area. Dr. Vladimir Olchenskov said that the warning centers permit scientific observation of animals in natural or near-natural habitats. Some of the observation teams' reports would be hard to believe if they weren't so carefully documented. "Ants pick up their eggs and move out of ant hills in a mass migration before earthquakes... Pheasants chorus an alarm before earth tremors. Goats and antelopes refuse to go into indoor pens for months before earthquakes. Tigers and other big cats do the same a few weeks before quakes occur."

Olchenskov stated that

> these have not been isolated occurrences but are part of a far-reaching pattern in nature which our researchers out in the field have witnessed time and time again. There is no question but that animals have alarm systems which warn them to leave areas which are about to become dangerous to them... In far eastern Kamchatka, not a single bear was killed in the winter of 1955–56 when a powerful volcanic eruption devastated the area. Hundreds of people died but all the bears had actually left hibernation and gone to safer lairs—long before the volcanic activity....

* * *

Western interest in the possibility of animals as reliable earthquake forecasters was sparked by the strange tales brought back by scientists who visited China in 1974 and 1975. These scientists reported that Chinese government workers purposefully gathered reports from zoo keepers and farmers about animal behavior. For example, on the basis of snakes coming above ground and freezing in the winter weather, of pandas in the zoo doing a curious dance, and of strange behavior of livestock in the vicinity of Haiching, officials had the city evacuated. A few days later, on February 4, 1975, a strong earthquake (7.3) destroyed about 50 percent of the city. Tens of thousands of lives were saved. The government earthquake office has even printed an illustrated pamphlet advising peasants of a wide range of erratic animal behavior that could mean that an earthquake is imminent. The August 9, 1976, issue of *Time* reprinted a portion of this pamphlet; it includes these signs:

> Cattle, sheep, or horses refusing to get into the corral.
>
> Rats running from their hiding places.
>
> Chickens flying up into trees and pigs breaking out from their pens.
>
> Ducks refusing to go to the water and dogs barking for no obvious reason.
>
> Snakes coming out from their winter hibernation.
>
> Pigeons in a frightened state not returning to their nests.
>
> Rabbits, with their ears standing, jumping up or crashing into things.
>
> Fish jumping out of the water as if frightened.

Based on the success of the Chinese, many American researchers were won over. On November 25, 1976, Dr. Barry Raleigh of the USGS in Menlo Park, California, told a *Washington Post* reporter that "there appears little doubt that animal behavior strongly influenced their prediction." Dr. Bernard Bodiansky of Harvard University exclaimed, "The animals did it!" Dr. Frank Press, of the Department of Earth Science at the Massachusetts Institute of Technology and a member of the National Science Board, told the same reporter, "Animal behavior as a means of predicting earthquakes must now be taken very seriously." All of a sudden reports of the animals in the Anchorage zoo acting strangely before the 1964 Alaska quake, and horses near Hollister, California, refusing to go into stalls before the 1974 Thanksgiving quake were remembered. Dr. Yash Aggarwal, the Columbia University researcher who was one of the group that discovered the "rock dilatancy" phenomenon that precedes quakes, is convinced. He now wants to see research on how long before a quake animals begin behaving strangely. Aggarwal told a *New York Daily News* reporter, "If it was even an hour, it would be possible to sound a warning so people in the area of the quake could take to the open. It's always safer in the open."

In response to this cry for "animal-earthquake" research, in the spring of 1977 scientists built homes for pocket mice (below ground) and for kangaroo rats (above ground) so they could join the cockroaches already living in refrigerated homes in their vigil along the San Andreas Fault. Normally mice are active at night and the rats are active during the day. The USGS has electronic devices monitoring all their behavior.

In another response to the cry for "animal-earthquake" research, in October, 1976, seismologists, geologists, and biologists met under USGS sponsorship at Menlo Park, California, to discuss the subject. A 429-page report on the conference and animals' quake-sensing abilities was released. But there was no mention of man's abilities in this area. If chimpanzees can sense quakes, why not man? Why not psychics?

Dr. Clarence Allen, a geologist at the California Institute of Technology, notes that "there are a lot more animals in the world than there are going to be instruments." But scientists continue to hope to develop reliable quake-predicting instruments—instruments at least as reliable as animals are proving to be. Scientists are now trying to discover what animals use to pick up earthquakes before they occur. If this can be done, the reasoning goes, they could build an instrument to do the same thing. Allen points out that we already know, for example, that dogs can hear better than people and birds can feel extremely slight vibrations. Dr. Jack F. Evernden, the USGS geophysicist who compiled the animal/earthquake conference report, believes that animals may sense changes in the earth's magnetic field before earthquakes. Thus far, however, no physical sensory mechanism has been identified that is common to all animals that show erratic behavior before earthquakes.

What might cockroaches, mice, pandas, horses, birds, apes, and other animals have in common that would enable them to sense impending earthquakes? The USGS would do well to read the research reports from parapsychologists who have tested the ESP in animals. Extensive studies have even been made on cockroaches and mice—two of the USGS's favorites. ESP, rather than some physical sense, may be the most fruitful area of research on this topic.

The research up to 1974 is usefully summarized by Dr. Robert Morris, the research coordinator of the Psychical Research Foundation, Durham, North Carolina—himself an investigator of psychic functioning in animals and humans—in his chapter in Apollo 14 astronaut Edgar Mitchell's *Psychic Exploration*. To clear away misconceptions on animal psi, Morris points out that the homing ability of many animals, including dogs, cats, and birds who have found their owners many hundreds of miles away, is not conclusive evidence of psychic ability. Nor are the cases of clever animals that appear to talk, do arithmetic, or read. In fact, some of the latter have been known to be due to training that depends upon the presence of a human who cues the animal, deliberately or otherwise.

It is in precognition—the ability to foretell events beyond analysis and logical reasoning—that animals appear to be genuinely gifted. This is precisely where we need assistance in earthquake research. According to Morris, "the strongest and most consistent set of animals studies of which we are presently aware involves mice, gerbils and hamsters in a two-choice precognition procedure." He goes on: "Evidence for psi seems obtainable from a wide range of species and central nervous system complexity levels."

Beyond this there are many anecdotal stories of animal ESP, ranging from pets howling upon the death of their far-distant human owners to pets answering questions that investigators themselves didn't yet have answers to. My favorite story, as reported in *Psychic* (September/October, 1973), concerns Missie, a Boston terrier, who astounded numbers of people with her ability in the 1950s. Notarized affidavits from radio moderators, superintendents of schools, veterinarians, and government officials attest to Missie's psychic ability. Here are two:

> To Whom It May Concern:
>
> When I was moderator on a radio talk show on station KTLN in Denver, Colorado, I had a phone call from Mildred Probert on 9/30/65 (the day my baby girl was born). Miss Probert told me her "psychic dog," a Boston terrier, had been barking out a yes answer when asked if my baby would be a girl.
>
> This call was made before the baby was born.
>
> She put the dog on the phone to bark out the hour of time it was then and the temperature (which I checked with a phone temperature call).
>
> Missie also gave the date, Miss Probert asking, "What month, date, year and the day of the week?" And how many letters in my name? All without error. It brought a rush of calls when I remarked it

was "the first time I talked to a dog and it answered back."

After that Missie performed over the phone on my program seven or eight times, giving scores for forthcoming football games and the World Series baseball games, correctly.

On New Year's Eve 1966, she gave the answers for events occurring each month for that next year. All turned out to be true.

Gary Robertson

To Whom It May Concern:

One day in the Spring of 1966, while visiting Miss Mildred Probert, her little Boston terrier, named My Wee Missie, gave quite a performance for me.

To my amazement the little dog, when asked by Miss Probert, barked out correctly my Social Security number, my phone number and address and the number of letters in the street on which I live. She then gave the complete birthdate, month and year.

She responded without hesitation and Miss Probert gave her no clues of any sort. Miss Probert would not have known these numbered items mentioned above. I can only respond to all this in much the same way someone in Shakespeare's play *Hamlet*, Act I, Scene V, says to Horatio, "There are more things in heaven and earth, Horatio, than are dreamt of in your philosophy."

Respectfully submitted,
Dennis Gallagher
House of Representatives
The State of Colorado
Denver

* * *

Ironically, the investigation of the ability of animals to sense earthquakes, now scientifically in vogue, seems to bring us back full circle to psychics and their ESP ability to predict earthquakes.

CHAPTER 11

Earth Changes and the Fulfillment of Bible Prophecy

"The earth is utterly broken down, the earth is utterly moved, the earth is staggering exceedingly.

"The earth shall reel to and fro like a drunkard and shall be shaken like a booth, and its transgression shall be heavy upon it, and it shall fall and not rise again.

"Then the moon shall be confounded and the sun ashamed, for the Lord of hosts shall reign. . . ."

Isaiah 24:19, 20, 23

"Tell us when these things will happen and what is the sign of your coming? . . ."

Jesus answered saying to them . . .

"For nation will rise against nation and kingdom against kingdom; and there will be famines and plagues and earthquakes in different places.

"Behold, I have told you in advance."

"Immediately after the suffering of those days the sun will be darkened and the moon will not give its light and the stars will fall from the sky and the power of the universe will be shaken."

Matthew 24:3, 7, 25, 29

"And there will be great earthquakes in different places, and famines and plagues; and there will be alarming sights and great signs will appear from heaven; and the winters will be severe.

"And there will be signs in the sun, and moon and stars; and on earth distress of the nations and confusion because of the roaring of the sea."

Luke 21:11, 25

(All quotations are from George M. Lamsa's *The Holy Bible from Ancient Eastern Manuscripts*, a modern translation directly from the Aramaic.)

What do Judaism, Christianity, some American Indian teachings, and many other religious philosophies have in common? They all assert that a messiah, a world savior, will come to earth. Most also prophesy that a specific set of events will precede this greatest of events. The predicted earthquake generation might constitute the period in which these prophecies are fulfilled.

The various events projected by psychics, geoscientists, meteorologists, astronomers, and others, as discussed in previous chapters, parallel the events prophesied by these various religious groups. In fact, there is a good degree of consensus among the often disputatious community of Christian interpreters of biblical prophecy that we are now entering the time of trouble forecast in the Bible. A good percentage of the American public agrees with these Bible scholars. More than ten million people bought Hal Lindsey's book *The Late Great Planet Earth*, which takes up these prophecies.

Lindsey, a graduate of Dallas Theological Seminary in Texas, is especially concerned in *The Late Great Planet Earth* with the prophecy of "... nation will rise against nation and kingdom against kingdom..." and the tribulations that will be suffered before the coming of the messiah to the Holy Land, which (as a Christian) he understands will be the return of Jesus Christ to earth. Lindsey bases his arguments on the Old Testament prophets Isaiah, Jeremiah, Ezekiel, Daniel, Joel, and Zechariah, and on Matthew, Luke, and Revelation in the New Testament. Lindsey says the formation of the state of Israel in 1948, with Jews from around the world returning to the Holy Land, is the fulfillment of the key prophecy of Jeremiah. This return is seen as a miracle. And when Jerusalem itself came under Israeli control as a result of the June, 1967, Six-Day War, the countdown officially began, in Lindsey's view. The formation of the European Common Market was seen by him as the next key step. In 1975 the nine-nation market endorsed Greece's application to become number ten. Lindsey sees the Common Market as the ten-horned beast described in Daniel and, with similar imagery, in Revelation.

Lindsey sees World War III breaking out in the Near East as the Russians, Egyptians (leading an Arab confederacy), China, and the African nations, all under the banner of communism, attack Israel. Lindsey then points out how the siege of Jerusalem by all these nations is prophesied in Zechariah, and a description of the battle given. In Lindsey's sequence, the Russians double-cross the Arab nations, but then the Russian army is destroyed by "fire" (Ezekiel), which Lindsey says is a nuclear attack. According to Revelation, the final clash takes place at Armageddon, the valley in Israel that contains the ancient Hebrew town of Megido. In this battle, according to Lindsey, a mainly Chinese army millions strong will threaten to annihilate Israel. This in turn sets up a confrontation in the valley between the army of the Chinese and the army of the Common Market nations, as led by a charismatic leader called the Antichrist and his False Prophet.

Lindsey says that Armageddon will be the greatest battle of all time. A frightful carnage will result, so great that the human mind cannot conceive of it. In the midst of this battle, in accordance with Revelation 16:18, an enormous earthquake will hit: "...there was a great earthquake, such as was not since men were upon the earth, so mighty an earthquake and so great." The shock wave from this earthquake and battle is supposed to produce worldwide destruction. As a result of this earthquake and/or nuclear blasts from the battle, the earth will shift on its axis. (Remember that Chapter 7 pointed out that some geologists maintain that a nuclear blast could set off the polar ice surge that unbalances the globe.) Finally, Lindsey says, at the very moment when all looks darkest, Jesus will return and save humanity. Lindsey believes that this series of natural and political catastrophes is primarily designed to shock people into accepting Jesus as messiah.

Lindsey's argument has relevance for Jews as well as Christians. According to Jewish teachings, the messiah is yet to come. The fifth-century prophet Zechariah not only refers to a great battle (Armageddon?) but also to the messiah then coming forth and to an earthquake and pole shift, the very same sequence Lindsey projects. Zechariah 14:4 says that the quake shall split in two the Mount of Olives outside Jerusalem, producing a great valley. In another passage (14:6), pole shift is indicated by: "And it shall come to pass in that day, that there shall be no light but cold ice." This fits with the earlier prophet Isaiah's prophecy of a large quake and pole shift (the relevant passages were quoted in the beginning of this chapter). In still other passages (19:5–6) Isaiah makes reference to the destruction of Egypt and a drought preceding the great quake. It is remarkable how the same events, in the same sequence (as discussed in the previous chapters of this book), are all foretold. The pattern sounds all too familiar.

In Luke 21:11 we get increasingly severe winters. These are, in fact, presently occurring as we move toward another ice age. In Isaiah 19:5–6 we get drought. In Chapter 6 we learned the connection between severe winters and droughts. Recalling the CIA report mentioned in Chapter 6, droughts and severe winters would certainly produce the famines and plagues foretold in Matthew 24:7 and Luke 21:11. The relationship between droughts, the water table, and earthquakes has been noted in Chapter 5. Earthquakes occurring around the world are prophesied in Matthew 24:7 and Luke 21:11, which would be the equivalent of the earthquake generation being foretold by the psychics. Isaiah 24: 19–20, 23 and Revelation 16:18 both foretell a quake of unprecedented proportions. Revelation 16:19–20 says this quake will devastate all the major cities of the world and make islands and mountains disappear. Luke 21:25 foretells the sea roaring, presumably over the land. We have seen in Chapter 4 that such huge quakes have occurred in the distant past. And remembering in Chapter 7 how Drs. Smylie and Mansinha found the poles to move as a result of large earthquakes, it is not surprising that Isaiah 24:19–20 speaks of the earth being "moved," "staggered," "reeling to and fro," and falling as a result of this great quake. Zechariah 14:6, Matthew 24:29, Luke 21:11, 25–26, and Revelation 16:21 all describe phenomena that indicates a pole shift, from the stars appearing to fall from the sky, to the movement of the sun and moon being interfered with, to sudden cold, ice, hail, and climatic change, to other "alarming sights."

Cayce and my psychic group feel that the biggest change the earthquake generation will experience will be spiritual—not in the form of organized religion but in *awareness, attitudes, and concern for others*. In a life reading for a friend of mine, Abrahamsen talked about veils being lifted from people's minds during this period. As these veils are lifted, people will understand better who they are, where they have come from, and what their fundamental spiritual nature is.

An entirely new vibrational environment is projected for the earth. In addition to the spiritual changes people initiate, they will become more spiritually aware as a result of the geological changes foreseen to occur because these changes eventually will establish a more harmonious vibrational environment to live in. The earth will undergo a self-cleansing process to overcome the distorted vibrations that man, through his free will has created over the centuries. (This correlates, incidentally, with the American Indian prophecies of a time of Great Cleansing.)

The psychics particularly pointed out the failings of the United States, with its environmental pollution, governmental corruption, judicial abuses, shortcomings of organized religion, deterioration of city life, disintegration of the family structure, rising divorce rate, deterioration of personal ethics, discrimination against minorities, and the extreme materialism of science and commerce. The United States, the psychics said, was overcome by insecurity, fear, and greed. They felt that the nation had substantially turned away from the original principles and goals of the founding fathers and that the geological upsurge would halt the United States' unbridled technology and materialism, for the safety of all mankind.

My psychic group felt that each of us would personally be forewarned through dreams and visions of the catastrophic events to occur. They recommended prayer and meditation in order to help our "channels" open. Like Matthew, though, they feel the exact day and hour of the great earthquake and the return of the Christ is not predictable.

Frank Waters, in *The Book of the Hopi*, says that the final warning before the cataclysmic changes the Hopi believe will occur before the emergence of the Fifth World will come when the Saquasohuh (Blue Star) kachina dances in the village plaza. Sasquasohuh represents a blue star, far off and not yet visible, which will make its appearance soon. The kachinas of the Hopi represent the invisible forces of life and Hopi friends have told me that the appearance of Sasquasohuh will be the real thing, not just a man dressed as this kachina. Waters says the time beginning the cataclysmic changes is also foretold by a song sung during Wuwuchim, the first ceremony in the Hopi's yearly ceremonial cycle. This song was sung in 1914 just before World War I and again in 1940 soon after World War II began. The Hopi also believe that plant forms from previous worlds are beginning to spring up as a sign of the changes to come.

The question now is: what should those of us who live in the high-seismic-risk areas do once we receive sufficient warning? Recently, I was part of a panel of researchers in parapsychology. Someone in the audience asked a panel member who lived in California what he thought about living there. The panelist was familiar with the Cayce readings. In fact, he had guided his life on the basis of many of them and had written a best-selling book about these readings. The panelist answered that he wasn't about to move and that if the Good Lord wanted him to move he would, but not until then. This seems consistent with the biblical attitude of trusting to receive the type of guidance we need. But, considering this person's strong attitude about not moving, I wonder if he would pick up any subtle guidance in the first place. As for a divine message, I wonder if he really needs one. After all, here is someone already convinced, and acting on many of the other insights of the Cayce readings.

This instance reminded me of when Jesus was tempted by Satan to throw himself off the top of the temple. Satan told Jesus that God's angels would surely keep him from injury. "Jesus said to him, 'Again it is written that you shall not tempt the Lord your God'" (Matthew 4:5–7). If one has already received sufficient warning—as this panelist obviously had, since he had based his life on Cayce—why tempt (i.e., test) the Lord further? As far as what constitutes sufficient warning, a life reading by Cayce gives one graphic instance. Cayce told a person that he led a righteous life in Atlantis but still went down with it. The person asked why, if he was so righteous, he had died in the catastrophe. Cayce's answer was simple: "You were warned." This may be an illustration of St. Peter's admonition to add *knowledge* to faith.

Most important, my psychic group said that spiritual help for man is on its way. Cayce said the last of this century ". . . will be proclaimed as the period when His light will be seen again in the clouds" (396-15, January 19, 1934). Once Cayce was asked what is meant by "the day of the Lord is near at hand." His response was:

> That as has been promised through the prophets and the sages of old, the time—and half time—has been and is being fulfilled *in this day and generation*, and that soon there will again appear in the earth that one through whom many will be called to meet those that are preparing the way for

His day in the earth. The Lord, then, will come, "even as ye have seen him go." (262-49) (emphasis added).

Abrahamsen, Karish, and Elkins also speak movingly of the coming of a messiah. They all speak of the return of "John the Forerunner" and of Jesus Christ. According to Cayce, by the year 2000 a number of Christ's disciples—Andrew, Bartholemew, Jude, Judas, John, and Peter—will be living on the earth again. Then John the Baptist will appear, going by the name of John Pineal, and give to the world the new order of things before Christ finally appears. According to Abrahamsen, Christ will bring many spiritual teachers with him. Abrahamsen said that people from outer space will also come to the earth, primarily to observe. He said they are too afraid of what our reactions to them would be to make themselves fully known at this time. Abrahamsen said that while these space people are spiritually evolved, they are not nearly so evolved as those persons who will return with Christ—people who won't need any spaceships. Frank Waters writes that Hopi prophecy says the coming Fifth World is the world of Taiowa the Creator and his nephew Sotuknang.

If the earthquake generation actually comes to pass, what of those who die? The psychics, who all believe in reincarnation, said that those who perish won't really have lost their lives. They will have many opportunities to reincarnate and live out new lives on earth. A friend of mine, Oswald White Bear Fredericks, the Hopi who supplied the source material for Frank Waters' book, told me that those who lose their lives will wait out a few worlds while Taiowa and his nephew Sotuknang teach the people and help them evolve spiritually. Then those who lost their lives will have an opportunity to return to earth and continue to evolve with the aid of those who have recently been taught.

What of the survivors of a possible earthquake generation? From psychological studies of the survivors of tragedies, we have learned that they feel better if they think the tragedy served a purpose; otherwise they are filled with anger, grief, and guilt that they survived. In the case of the 1977 Kentucky nightclub fire that took so many lives, the survivors felt relieved when better safety regulations were instituted. Cayce and other psychics feel there is a definite plan for the survivors. They have spoken in mystical terms about a new "root race" coming into existence with the Aquarian Age. They have talked of small communities springing up that practice new life-styles based on new concepts of consciousness. The Hopi speak of the humble people of little nations, tribes, and racial minorities leading this transition. The depression of the 1930s was a great leveler. Likewise, if an earthquake generation occurs, it seems that people will all start out again on an equal basis.

The psychics believe that nothing happens by chance and that the survivors will have the opportunity for true spiritual growth. The survivors will learn to give thanks in the midst of catastrophe. The greatest growth will come to the children as they learn to believe in the presence of God and to contact "higher" sources for information and guidance. But the psychics note that there will be dogmas and that humanity will have to learn how to stop enslaving itself by its own laws. They say that humanity will have to learn to make practical use of all the knowledge it has accumulated and the resources it has been given. According to the psychics, since the United States in particular has not done so well in these areas, and has not cared for all those in need, it will undergo the various destructions foreseen and will have the opportunity to learn how to handle fewer resources.

What can be expected in this new order of things, the millennium that is supposed to follow Christ's return? Cayce said that we should all pray to be incarnated during this time. But this golden age is not supposed to happen instantly. It will take several hundred years to develop fully. I asked Abrahamsen to give me a brief account of specific changes. He told me that government, instead of regulating people, would put its emphasis on helping people to develop and regulate themselves. Mind-to-mind communication will be commonplace. While transportation will still be dominated by the airplane, there will be some teleportation, à la "Star Trek." Education will teach people how to tap their inner power. Agriculture will use prayer to control crops and rainfall. The sun and the earth's electromagnetic field will be the main sources of energy. Psychics will be used to guide successful research. Medicine will rely heavily on healing via color, often with dramatic results. Cancer will be cured. There will even be limb regeneration. Crystals are also supposed to be used for healing and energy.

Taking a long-range view, it seems that despite the threat of disaster from cataclysms, the earthquake generation has a great deal to look forward to.

CHAPTER 12

Coping with Reality: The Problems, the Programs

> "When a scientist states that something is possible, he is almost certainly right. When he states that something is impossible, he is probably wrong."
>
> Arthur C. Clarke

> ". . . experts have noted that the quake problem seems too huge, the individual too powerless and the likelihood of harm too remote to keep quakes in the forefront of human worry."
>
> Thomas Canby and James Blair,
> "California's San Andreas
> Fault," *National Geographic*,
> January, 1973

> "There are, it seems, a lot of fools now living in California."
>
> Dr. John Gribbin and
> Dr. Stephen Plagemann, in
> *The Jupiter Effect*

On a visit to New York City, where I grew up, I stopped to have Antonio, my old barber, cut my hair again. As I sat in the chair, I talked to him about the earthquake book I was writing. I ran through a lot of facts about quakes, but as soon as I began to tell him about the great number of earthquake casualties in the last century and other gruesome statistics, Antonio stopped cutting my hair. Tears began to well in his eyes. He quietly said, in his strong accent, "Don't tell me about if earthquakes are real—I tell you." Antonio then told me how, when he was a boy in Greece, visiting the outer Greek isles, a moderate-sized quake struck the mainland. He took the first boat back to check on his family. As he rushed to his village, an aftershock hit and he almost was crushed beneath a cascade of mortar and bricks. But he didn't dramatize his close call nor go into statistics. In a strained voice he simply said that his father and older brother were killed in the initial shock. Nameless statistics are not simply the dead—they are our loved ones. Antonio carries a dread of earthquakes with him. Some of the movie stars whose hair he cuts have repeatedly tried to have him move his shop to California, but he'll have no part of it. I didn't have the heart to tell Antonio about the faults that lie beneath New York City.

It may be unfortunate that for the rest of us, the threat of earthquakes is not real even though almost two-thirds of the American people live under it. The 750,000 Chinese who lost their lives in Tangshan in 1976 are seen as the victims of a freak accident. But we fail to realize that earthquakes are not freak accidents. Earthquakes can be expected in China, just as they can be expected in California. They are part of the earth's natural processes, yet we have built major cities in the heart of some of the earth's most active seismic areas. The public must recognize this danger and adjust to it, regardless of whether the psychic prediction of a generation of particularly concentrated earthquake activity is correct. Even if quake activity continues at no more than its present level, the threat is enormous.

For example, in the San Francisco Bay area 4.7 million people live in double jeopardy, between the Hayward and San Andreas faults. The U.S. Office of Emergency Preparedness (OEP) has put forth several scenarios based on an 8.3 quake (equivalent to that of 1906) striking the area. The first has the quake striking at 2:30 A.M., while most people are asleep at home. The toll: 3,000 killed, 11,000 hospitalized. In the second version, the quake strikes at 2 P.M., while most people are at work. The OEP sees 9,500 killed and 35,000 hospitalized. The third

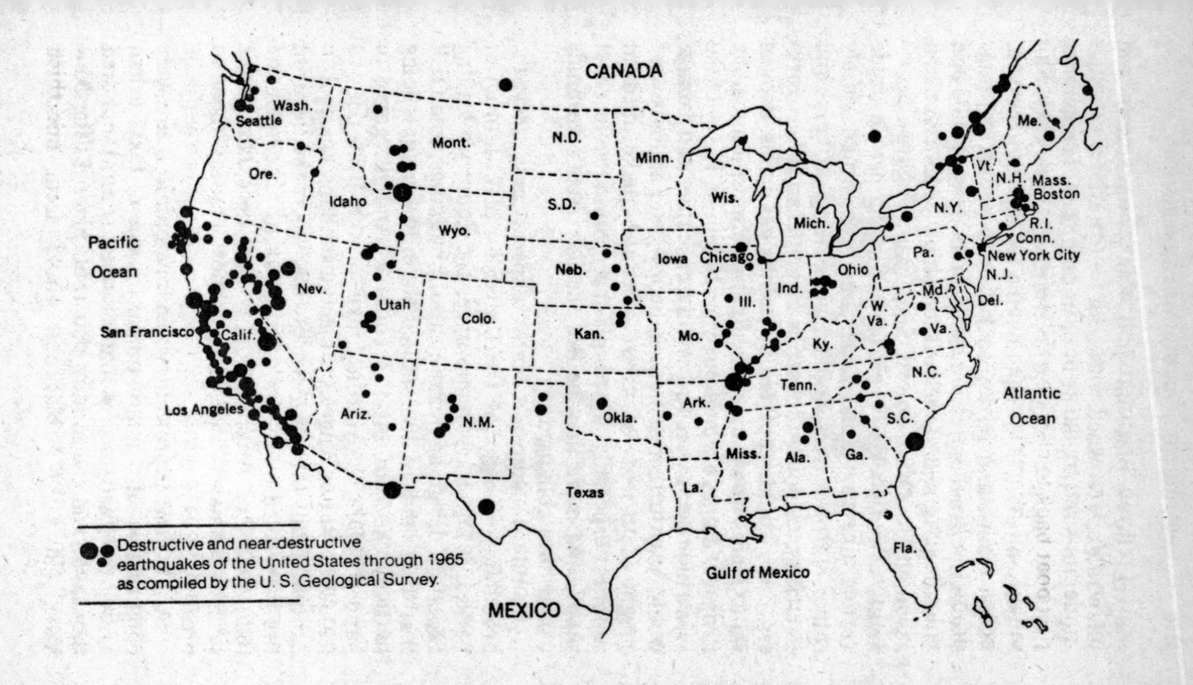

Destructive and near-destructive earthquakes of the United States through 1965 as compiled by the U. S. Geological Survey.

scenario takes place at 4:30 P.M. while roads and sidewalks are crowded with rush hour traffic. Casualties could zoom to 100,000 dead or injured. *Time* (September 1, 1975) painted a grim picture of "The Day San Francisco Is Hit." One can imagine, as *Time* did, the cars, buses, trains, and ferries packed with people, as huge cracks open in the streets and sidewalks. There would be a rain of glass and stonework crushing pedestrians and automobiles. Office towers and luxury highrise apartments would rock to their limit, flinging people inside around like rag dolls. Older buildings would simply collapse. Power lines would snap, cutting off lights and electricity, trapping thousands of panicked office workers. Fires would erupt, fed by leaking gas mains. Broken water mains would make fire fighting almost impossible. Property damage would be over $10 billion. If any of the area's nine dams were to give way, casualties and damage would soar further as houses, people, and animals were caught in torrents of water. After the quake, medical supplies, especially blood plasma, would be in short supply and more than half of the Bay Area's hospitals would have collapsed.

Could it happen? Many experts now feel the probability is high. At the University of California at Berkeley, Dr. Karl Steinbrugge, the country's leading expert on designing quake-resistant buildings, told OEP that "thousands of lives snuffed out in 30 seconds is going to blow the roof off this country. And it's going to happen!" Dr. C. Martin Duke of the University of California at Los Angeles projects a similar scenario if an 8.3 quake hits the Los Angeles area, where the threat of fire is replaced by that of dams breaking. The moderate (6.5) 1971 Los Angeles County San Fernando quake severely damaged two dams, almost causing them to break.

All of these projections, in my opinion, are very conservative. If a major quake hit one of these urban areas, the casualties would number several hundred thousand, and the looting that took place during New York City's 1977 blackout would seem minor in comparison. Added to the physical damage would be the social and psychological damage. After the 1971 San Fernando quake, schoolchildren were reluctant to return to school. Here was something neither mommy or daddy could control, something that made them fearful. Many children and adults had the "earthquake jitters" for months. The common denominator of the symptoms was anxiety triggered by the quite realistic fear of a quake's havoc. Geography professor Harold D. Foster of the University of Victoria, British Columbia, produced a disaster magnitude scale that was based on (1) psychological and social disruption, (2) injuries and deaths, and (3) environmental damage with which survivors must cope. Under his criteria (which do not distinguish between man-made and natural disasters) World War II ranked the highest at 11.1 and the 1923 quake that devastated Tokyo was ranked fifth at 9.1. The 1976 Tangshan quake received a score of 9.0. Surprisingly, the atomic bombing of Hiroshima only received a score of 8.2. The ranking of earthquakes ahead of atomic bombings serves to illustrate what total disasters earthquakes can be.

How does one draw the public's attention to this serious subject, especially if Hugh Auchincloss Brown and others calling for a capsizing earth are evenly remotely right? The public turned out in droves to see Charlton Heston in *Earthquake*. The British Broadcasting Corporation produced a documentary film entitled "The City That Waits to Die" that focused on San Francisco's peril. It was also shown in the U.S. on the Public Broadcasting television network. Yet no one is terribly concerned, and certainly not the U.S. government which, according to Dr. Frank Press of the Massachusetts Institute of Technology, budgets "a ridiculously small" amount of money for research and communication on the subject.

Allen Eckert based his chilling novel of global disaster, *The HAB Theory*, on Hugh Auchincloss Brown's life and work. In it, the fictional Herbert Allen Boardman tries for more than half a century to get people in the universities and government to listen to his ideas (the HAB theory), even going so far as publishing his own book. Finally Boardman, old and tired, tries one last and desperate way to gain attention for his theory. Boardman plans to shoot the President, and he does—but only with nonlethal wax bullets. Thus Boardman finally captures attention from both the President and the public.

Here is the real obstacle—the problem of getting people to consider views different from their own. In diversity lies the key to human progress and survival. Science and the passage of time have proved many theories wrong that man has held dear. But if we test a number of different theories, the price for error will not be so great. We won't find ourselves way out on a limb that is about to be cut off. Should an earthquake generation be upon us, there is still time to make adjustments. If the earth does undergo cataclysmic change, but we ignore the warning signs because of our faith in an imperceptive scientific establishment, how will that establishment make amends for the many deaths likely to occur? How many scientists–opinion-makers are capable of dealing with such diverse fields as geology, meteorology, astronomy, archaeology, and biology, which all seem to indicate the possibility of an earthquake generation? If the key warning sequence given by psychics begins, there will still be time to avert the worst consequences of further heightened geological activity. If the temperature predictive factor is proved correct, this alone justifies the effort of considering the psychic viewpoint. If a psychic can occasionally pinpoint an earthquake—as Clarisa Bernhardt has done—why not closely monitor his or her predictions as part of a multifactor early-warning system? Since psychic research has shown that forewarnings of disaster show up in dreams also, why not develop a public registry for such dreams, just as there are now several private organizations recording premonitions? Psychics can aid science if scientists will drop their prejudices and anxieties about what psychic ability implies for their oftentimes smug view of reality.

On December 16, 1977, the federal government announced through the National Earthquake Information Service (NEIS), a division of the USGS, that it had allocated $5,000 to test the accuracy of psychics in predicting earthquakes. NEIS had earlier recorded some of Clarisa Bernhardt's remarkably accurate predictions of the day, time, place, and magnitude of earthquakes—(at the moment, her record stands at seven out of seven). Dr. John Derr, coordinator of NEIS, is the man who took Clarisa's predictions and spoke so highly of her accuracy. The newly announced project is headed by Dr. Roger Hunter, a geophysicist who works only part time on the project from an office on the campus of the Colorado School of Mines in Golden, Colorado. Hunter, in a statement to the press, said that "persons who believe they have the power to forecast earthquakes should mail us the data and we'll feed it to the computer... This program doesn't act against anyone with real abilities. Actually, it is the best possible way to prove such an ability."

When I read this announcement I was aghast at the naiveté of the program. While I don't doubt that it is well-intended, I must nevertheless ask: what makes the National Earthquake Information Service think that the best psychics will come forth? What makes NEIS think that its way is "the best possible way to prove such an ability"? What makes Hunter, a geophysicist, think he has any idea of how to work with a psychic and test his or her ability? What makes the NEIS confident that a meager $5,000 study will in any way touch on the potential of psychics to give earthquake predictions? Quite frankly, the program seems designed to avoid meaningful research, or at last to fumble the ball if it gets any useful data. I feel sure that NEIS's ignorance of the physiological, psychological, and methodological variables involved in working with psychics and obtaining useful information will ensure that.

There is a personal connection in all this because I am an alumnus of Colorado School of Mines, where Hunter has his office. After I graduated and went on to do graduate work, I spent a number of years working with psychics and researching their abilities. I have worked with almost fifty psychics who were screened from many hundreds with some degree of psychic ability. More than a thousand psychics—laboratory-validated or self-styled—have mailed information to me. Some of the psychics I worked with have demonstrated incredible abilities, abilities that can be developed like any other. My discovery of the Flagstaff site through psychic means is an example. But if I have learned anything in this work, it is that the procedure the NEIS is employing to test psychics is not likely to prove much about psychic ability.

I sincerely suggest that the NEIS or some other agency take out an "insurance policy" of sorts by allocating a realistic amount of money to properly evaluate psychic abilities and information which psychics with good "track records" produce, no matter how controversial or "far out" it seems. At present, the gamesmanship involved in receiving research money prohibits top minds from investigating such areas. The well-known secret is that only those who think the way grant committees do get money for research, and psychic research is at the bottom of the list—if it's there at all! In *The HAB Theory* one scientist says,

> Most of us as scientists, Mr. President, encounter many riddles within our own fields, riddles for which there are no explanations, however simple or complex. Such riddles bother us enormously, though we seldom allow this to show. Deep inside, though, we wish we had the time to really devote ourselves to exploring the mysteries.

Scientists and governmental authorities must not only recognize their responsibility to the public, they must also recognize their responsibility to themselves as well—and get on with researching the mysteries.

In addition, the media must recognize their responsibility to report accurately and extensively in this area. They should report the diversity of views on any scientific subject and investigate them in depth without prejudice. When the science editor of a major newspaper can follow the technicalities of a scientific discussion and then align himself with the scientific establishment against all diverse views, he is not doing his job of reporting. How many creative thinkers does society produce in the class of Galileo, Newton, or Einstein, men with the ability to see far beyond established truths? What was their "press" when they first proposed their ideas? And how are we to learn of their work and others of their potential if not by the press?

Just how are we to cope with the threat of severe earthquake activity? When Dr. Willard Brinkman, a professor at the University of Wisconsin, tried to set up a conference on "natural disasters and people's attitudes towards them," no one showed any interest. And here lies the major problem in dealing with earthquake survival: apathy. Unlike us, the Chinese have taken unprecedented action in regard to the threat they face. The entire country has become involved. They have set up a nationwide prediction and warning system with ten thousand special workers manning observation posts. In addition, countless peasants, teachers, students, and others aid in the program. According to a team of ten American specialists who toured China in 1976, the Chinese effort "encompasses *every* prediction method that has ever been suggested in any part of the world." The Chinese say that they have already successfully predicted eleven major quakes, evacuated cities, and saved tens of thousands of lives. Even before the mighty Tangshan quake that took an estimated 750,000 lives—the most notable and unfortunate miss for the Chinese warning system—*Science News* reports, "some last minute, unequivocal harbingers were manifest, but there was too little time to process the information and issue a warning." This has led the Chinese to develop their warning system even further.

On the other hand, Americans are doing little to cope with the earthquake threat. Dr. Clarence Allen of the California Institute of Technology points out that "the pressures of population growth are causing expansion into areas that are more difficult to develop safely than those of past decades—often into mountainous areas, active fault zones, or areas of artificial fill that necessarily have earthquake-related problems associated with them."

While the Chinese are moving from quake-prone areas or are willing to live in simple makeshift shelters, we are building subdivisions that straddle faults and skyscrapers that climb even higher in our threatened cities. Few building codes are following architects' recommendations for less glass and more steel cross girders for buildings or for homes designed to absorb or dampen shock waves. Few people are buying earthquake insurance. Dangerous, old, unreinforced buildings in cities are not being pulled down or being made safe. It is estimated that there are about fourteen thousand such buildings in the Los Angeles area alone and official figures say that about one hundred fifty thousand people either live or work in these buildings. Nor are dams regularly examined and made safe. A spokesman for the Los Angeles County Flood Control District's dam investigations section is reported (in a news release dated February 6, 1977) to have said that only two of the county's nineteen dams have been brought up to standard, while two others have been restricted for use because they need repairs. California's building codes are based on the assumption of an upper limit to the severity of earthquakes, and none is supposed to be much more severe than the one that hit Alaska in 1964. (8.3). Yet, as noted in Chapter 7, scientists propose that every fifty years a quake four times more severe is likely to occur. Dr. Charles Richter, the dean of the world's seismologists, told a San Francisco reporter, "There is simply no locality in the state [California] that is exempt from the earthquake risk. The probability of another major one is not a doomsday prophecy—it is going to happen."

A University of Arizona earthquake specialist, Dr. Charles Glass, reports that in the Salt Lake City area, where earthquakes have struck before, "buildings are going up along clearly defined fault lines. I don't think the buyers know what they might be up against."

Beyond discouraging building in earthquake-prone

areas and encouraging tougher building codes, state and federal governments might begin to set up "disaster brigades" which could come to the aid of the already established civil-defense network in the case of a city crippled by an earthquake. Retired military personnel and civilians could be organized and trained to fight fires, distribute food and water, build makeshift shelters, and give emergency first aid. Emergency evacuation networks could also be set up in case a quake warning is issued. Supplies must be available for people to eat and sleep outdoors for a period of time while "sitting out" a quake. Here problems of vandalism and looting of the evacuated area will have to be dealt with. It is also a good idea for families to put up long-term food supplies, especially if we have several years in a row of poor weather and bad crops, as has been happening on our march toward a new ice age.

In the final analysis, we all must make our own personal forecast of what the future may bring, and then we must make our own decisions on how best to cope with the outcome. And how we arrive at these decisions may be more important to the future of humanity than the decisions themselves. The potential threat of an earthquake generation is a call for *all* of us to rethink what our journey here on earth is all about.

APPENDIX A:

The Ten Most Dangerous Cities in the U.S.A., Plus Two

When the next major earthquake strikes—not *if*: that is the basis on which the President's Office of Science and Technology recently (6/78) announced a civil defense plan to cut losses of life and property. The ten metropolitan areas judged to be in greatest danger are San Francisco, Los Angeles, Salt Lake City-Ogden, Puget Sound (Seattle-Tacoma), Hawaii, St. Louis-Memphis, Anchorage-Fairbanks, Boston, Buffalo, and Charleston. (Due to recent discoveries, I believe New York City and San Diego should be added to the list.)

The report stated that an earthquake is by far the largest single-event natural hazard faced by the nation. And it also stated that we're in terrible shape to face one.

What makes each of the cities named in the government report places of high risk? To answer this question, let's look at the earthquake history of these particular cities.

San Francisco. The U.S. Office of Emergency Preparedness, noting that San Francisco Bay area residents live between the Hayward and San Andreas faults—a case of double jeopardy—has estimated that if a large quake struck along either of these faults during rush hour, more than 100,000 deaths or injuries could occur where fires and failing dams would make casualties soar.

The first significant quake in the San Francisco area occurred in 1800 and was recorded by the brothers at the then-recently-completed mission at San Juan Bautista. Damaging earthquakes occurred throughout the 19th century: in 1836, 1838, 1840, 1858, 1861, 1865, 1868, 1892, and 1897. The 1836, 1838, and 1865 quakes have been estimated to be about as large as any that have occurred in California.

The only large quake to occur in the San Francisco area in this century was the infamous 1906 quake, California's largest (approximately 8.3 magnitude). Damage to buildings was extensive in San Francisco, and, combined with the raging fires, over 500 million dollars in destruction ensued.

Ominously, each year minor-to-moderate earthquake activity takes place along portions of the San Andreas, Hayward, and Calaveras fault zones which pass under the Bay area.

Los Angeles. According to Dr. Charles Richter, the nation's foremost seismologist, "Los Angeles will undoubtedly suffer a maximum 8.5 magnitude earthquake soon." Besides the infamous San Andreas fault, L.A. has a number of other faults threatening it. For example, the Newport-Inglewood fault runs right through downtown L.A.

The first strong earthquake recorded for L.A. occurred in 1769. Two major shocks occurred in 1812: one killing 40 people attending church at San Juan Capistrano. There were significant shocks in 1852 and 1855. Then another major shock occurred in 1857 (estimated as equivalent to 8.0 on the Richter Scale) near Fort Tejon. An eyewitness reported that "Los Angeles was rocked from end to end, for two minutes or more... Houses rocked and toppled."

The L.A. area has not been hit by a major shock since the 1857 quake, but significant quakes occurred in 1893, 1899, 1918, 1920, 1925, 1933, 1941 (two), 1952, 1968, and 1971.

Researchers such as Richter have projected a figure of 200 years as average length of time between major earthquakes in southern California, and they agree that the 1971 quake didn't release the strain that has been building up in the area since the 1857 quake and thus the "inevitable" great Southern California quake may not be far away. (In fact, the 1971 quake struck in an area that has been inactive since the end of the last Ice Age—about 10,000 years ago.)

In fact, while all watch the Palmdale Bulge along the San Andreas fault sixty-five miles to the north, they forget that Robert M. Hamilton, Chief of the Office of Earthquake Studies for the U.S. Geological Survey, has discovered that Pasadena has dropped 5 to 6 inches in the past few years. Most ominously, Cal Tech scientist Dr. Karen McNally says that the pattern or swarms of small earthquakes (micro-tremors) now occurring in the area of the Palmdale bulge are similar to the clusters of microtremors now known to have preceded the 1971 San Fernando quake and the disastrous 1976 Tangshan, China, earthquake that took over 700,000 lives. McNally says that such swarms have preceded quakes from 2 to 10 years.

Salt Lake-Ogden. Few realize that Salt Lake City and Ogden straddle one of the most active seismic belts in the world. After years of studying earthquake activity at the base of the nearby Wasatch Mountains, Dr. Kenneth L. Cook, head of the University of Utah's Geophysics Department, concluded that the Wasatch fault is part of the World Rift System. California's infamous San Andreas fault would only be considered a small arm of the Wasatch rift. A detailed catalogue of Utah's earthquakes from 1850 through June 1965 (in the *Bulletin of the Seismological Society of America*, 1967) listed 609 events, many of them along this zone.

For example, three distinct shocks rocked the Ogden area on July 18, 1894, and cracked walls. On May 22, 1910, a moderate-sized quake (equivalent to approximately 6 on the Richter scale) damaged many chimneys and older buildings at Salt Lake.

The Ogden area was strongly shaken on May 13, 1914. According to the government report, "Windows were broken and chimneys were thrown down at Ogden; near panic was reported at Central Junior High School. The shock was felt from Collinston on the north to Riverton, south of Salt Lake City, an area covering about 20,700 square kilometers." Finally, the March 12, 1934, 6.6

magnitude shock that hit on the north shore of Great Salt Lake near Kosmo would have caused great damage had it centered in a more densely populated area. As it was, Salt Lake, though almost 100 miles away, still received moderate damage. With Salt Lake and Ogden sitting on top of the Wasatch fault, many geophysicists say is responsible for pushing California out into the Pacific Ocean a few centimeters each year, it is not hard to see how a really large quake will one day "inevitably" strike Salt Lake and Ogden.

Puget Sound (Seattle, Tacoma, Olympia, Vancouver): People watching for California's next quake tend to forget that the Puget Sound is also one of the most active seismic areas of the country. Most of the 850 earthquakes that were felt in Washington from 1847 to 1965 occurred in this region.

In recent times, it has been a region of frequent strong earthquake activity that has brought serious damage to Seattle, Tacoma, Olympia, and Vancouver, British Columbia.

The Puget Sound was struck by quakes of between 5.0 to 5.7 magnitude in 1872, 1909, 1939, and 1946. But a few months after the moderate February 1946 quake, a much stronger quake of 7.3 magnitude struck in the Strait of Georgia (just north of Vancouver) on June 23, 1946. This shock caused the bottom of Deep Bay to sink between 9 and 78 feet. Then, in 1949, a 7.0 magnitude quake struck near Olympia which caused more than 25 million dollars of damage. According to a government report, "at Olympia nearly all large buildings were damaged and water and gas mains were broken . . . railroad bridges south of Tacoma were thrown out of line . . . a sandy spit jutting into Puget Sound disappeared . . . a tremendous rockslide involving a half-mile section of a 270-foot cliff toppled into Puget Sound."

In 1965, a 6.5 shock struck dangerously close to Seattle and caused 12.5 million dollars in damage. Many chimneys in Seattle were damaged and in West Seattle, two schools were damaged, as were two brick school buildings in Issaquah (considerably).

Considering the frequency of strong quake activity in the area, it is easy to see how it is "inevitable" that a very strong quake will strike in one of the more densely populated areas.

Hawaii. The islands are considered threatened since they lie in both an active tectonic and volcanic area. Most of this seismic activity centers on the island of Hawaii because of its active volcanoes, but the strongest shocks are of tectonic origin. Shocks north of Hawaii are often felt strongly on Maui, Lanai, and Molakai.

The islands are also a frequent victim of earthquake generated sea waves or *tsunamiis*, often from far distant epicenters in the northwest Pacific and South America coasts. Of the 87 *tsunamiis* to hit the islands since the early 1800s, only 16 have caused significant damage, but only 5 of these originated near Hawaii. In 1946 following an earthquake in the Aleutian Islands, 55 foot high waves struck the north coast of Hawaii. At Hilo, hundreds were injured or killed and 488 buildings were demolished. Over 25 million dollars of damage was done. In 1957, another Aleutian island-generated *tsunami* did $3.0 million in damage, and the 1960 *tsunami* from a Chilean earthquake did $2.6 million in damage. *Tsunamiis* also did major damage in 1964 and 1975.

After the arrival of the missionaries in 1820, four or five earthquakes per year were reported during the 1800s. During the 1900s over 84 earthquakes strong enough to cause noticeable damage have occurred among the many hundreds that have been recorded.

The largest earthquake to occur in the islands hit near the south coast of Hawaii on April 2, 1868. It has been estimated that this quake would have been almost an eight on the Richter scale, i.e., almost as large as the infamous 1906 San Francisco quake. According to the *Earthquake History of the U.S.*, a government publication, this quake knocked nearly every wooden house off its foundation (at Keiawa, Penalua, and Ninole), and knocked down almost every wall in Hilo and caused landslides as far away as Waipo and Hamakau. Ground waves estimated at one to two feet from trough to crest were observed at Kohala. At Ulupalakua, Maui, the motion was so violent it was difficult to stand. The shock was distinctly felt over 350 miles away on Oahu and Kauai. Clocks stopped at Honolulu. A *tsunami* struck the Kaw-Puna coast and rolled in over the tops of the coconut trees; a height of at least 60 feet. Most houses were swept into the sea and a number of people drowned. At Hilo the height of the wave was 10 feet, at Kealakekua, 8 feet and at Koaluala 25 feet.

Severe earthquakes have also occurred: (1) in 1938 (6.8 magnitude), with an epicenter located under the ocean 40 miles east of the island of Maui. There was general panic on Maui as homes were damaged, and Olinda Reservoir was badly cracked; (2) in 1951 (6.9 magnitude), where cracks were opened in the coastal highway and scores of homes were wrecked or damaged on the west side of Hawaii, and (3) in 1975 (7.2 magnitude), on the southeastern coast of Hawaii on the south flank of Kilauea volcano. This quake set off a small eruption on Kilauea and generated a local *tsunami* that reached a maximum height of 45 feet and swept over a half mile inland. This recent quake caused about 4.9 million dollars of property damage. A quake in 1973 did over 5.6 million dollars in damage.

Most recently, in March 1977, volcanologists warned that Mauna Loa is swelling and that it could experience a major eruption by the end of 1979 that would threaten Hilo (30 miles away).

St. Louis-Memphis. Few people realize that St. Louis and Memphis of the Mississippi Valley lie within one of the world's most active seismic zones. A zone so strong that even the distant cities of Nashville, Louisville, and Cincinnati are placed in jeopardy. The record of earthquakes in the area prior to the nineteenth century is virtually nonexistent, but there is geologic evidence that the Mississippi Valley has had a long history of activity. Since 1811, at least 38 shocks strong enough to cause damage have occurred.

The three earthquakes which struck New Madrid, Missouri, from 1811 to 1812 were the most violent in U.S.

history, each larger than either the 1906 San Francisco or the 1964 Anchorage quakes. One resident of the then-sparsely populated area, caught by the first of the New Madrid quakes, wrote: "The whole land was moved and waved like the waves of the sea. With the explosion and bursting of the ground, large fissures were formed, some of which closed immediately, while others were of varying widths, as much as 30 feet." The Mississippi, thrown from its bed, swept away entire forests and great topographic changes occurred. Whole islands disappeared, and the Mississippi was rerouted, creating 10 mile long Reelfoot Lake. Chimneys were knocked down as far away as Louisville, Kentucky, and Cincinnati, Ohio, and buildings as far away as Chicago, Washington, D.C., and New Orleans were rattled. The main shocks were felt over an area of at least 5 million square miles and each shock would have measured over 8 on the Richter scale.

In 1843, another severe earthquake hit the New Madrid area and collapsed several buildings in Memphis. In 1895, nearby Charleston, Missouri, was hit with a shock that damaged every building in the commercial area of the city. The shock was felt over all or parts of 23 states and at some places in Canada.

The frequency of occurrence of earthquakes the size of the 1811–12, 1843, and 1895 quakes is low, but continuing minor to moderate seismic activity in the Mississippi Valley serves to remind us that in the manner typical of active seismic zones, we can expect large magnitude tremors to occur again. The moderate November 9, 1968, 5.5 shock that hit just south of St. Louis and the small March 29, 1972, 3.7 and the September 20, 1978, 3.5 shocks serve notice to this.

Anchorage-Fairbanks. More earthquakes occur in Alaska each year than in the other 49 states combined. As many as 4,000 are detected in a year. This is because the earth's most active seismis feature, the circum-Pacific belt, brushes Alaska and the Aleutian Islands.

The first earthquake recorded in Alaska occurred in 1788, and the absence of any large population during the 1800s let the major shocks of this time go virtually unnoticed. During the past 79 years, eight "great" earthquakes, earthquakes of 8 or more on the Richter scale, have occurred. Four of these caused topographic changes and extensive property damage (1899, 1900, 1957, and 1964).

In 1899, the Yakutat Range area of southeastern Alaska was struck by a quake which was rated 8.6 on the Richter scale. One eyewitness reported the shaking as "violent and impossible to stand without holding on to something." A 20-foot-high sea wave was set in motion and land in some areas was uplifted as much as 48 feet. In 1900, a 8.3 magnitude quake struck near Cape Yakataga in southeastern Alaska, and in 1957 a 8.3 magnitude quake caused very severe damage on Adak and Unimak Islands and generated a 40-foot-high wall of water which smashed the coastline. And Mount Vsevidof erupted after being dormant for 200 years. The *tsunamiis* this quake generated did over $50 million in damage.

Alaska's most notable quake occurred in 1964. This 8.5 shock devastated downtown Anchorage and left homes twisted and broken in the residential section of Turnagain. The resulting *tsunami* virtually destroyed many of Alaska's coastal towns and caused damage in Hawaii and Japan. All told over $500 million in damage was done. It was fortunate that this quake occurred on Good Friday, a holiday for schools, and a time when most people were out of office buildings.

The 1964 quake was the first time modern buildings in the U.S. were tested by an earthquake of above 8 magnitude (the last comparable shock occurring in San Francisco in 1906). Unfortunately, the multistory earthquake-resistive structures in Anchorage failed and suffered extensive damage.

While eight "Great" earthquakes have occurred since 1899, numerous "major" earthquakes (magnitude 7-7.9) have also centered in the State since 1899. While only 13 occurred in or near populated regions causing minor to severe damage, 150 or more have occurred in uninhabited areas. Among the more significant shocks: (1) a 7.3 quake hit south Fairbanks in 1937; (2) a 7.3 shock hit Fairbanks again in 1947; (3) a 7.3 shock hit central Alaska near Huslia on April 7, 1958; (4) a 7.9 shock hit southeastern Alaska in July of 1958 and (5) a 6.5 shock hit the central interior near Rampart in the Yukon River in 1968 and during the first 24 hours after this 1968 quake, over 2,000 aftershocks were recorded. It seems like Alaska never stops shaking for long.

Boston. Boston has a history of relatively rare but devastating quakes. Since Colonial times, 15 earthquakes strong enough to cause noticeable damage have centered in the greater Boston area. The first such quake was recorded in 1643 and the last in 1963. Several of the 1700 and 1800 quakes have been described as "tremendous" or "violent" and caused considerable damage. The moderate quakes of this century may be premonitory warnings of another "tremendous" quake to come.

But Boston's high risk classification most directly results from the devastating November 18, 1755, earthquake which struck at Cape Ann (outside of the greater Boston area). John Winthrop of Harvard College was one of the Boston colonists tumbled out of bed by this quake. He reported that "the earthquake began with a roaring noise in the northwest like thunder at a distance; and this grew fiercer, as the earthquake drew nearer." The quake threw down walls and chimneys and cracked wooden roof beams. According to a report by Carl A. von Hake of the National Oceanic and Atmospheric Administration which appeared in the U.S. Geological Survey's *Earthquake Information Bulletin*, there was "violent movement of the ground, like waves of the sea, making it necessary to cling to something to prevent being thrown to the ground. At Pembroke & Scituate small chasms opened in the earth through which fine sand reached the surface. Large numbers of fish were killed and many people on vessels felt shocks as if the ships were striking bottom. This earthquake was felt from Lake George, New York, to a point at sea 200 miles east of Cape Ann, and from Chesapeake Bay to the Annapolis River, Nova Scotia, about 300,000 square miles." It was fortunate that the buildings in Boston at this time were low and mostly of wooden frame construction. This 1755 quake would have measured about 7 on the Richter scale.

Buffalo. Few people realize that Buffalo, Rochester, and Massena lie along one of North America's most active seismic regions, a zone which follows the St. Lawrence and the south shore of Lake Ontario. Strong earthquakes occurred in the St. Lawrence Valley in 1638, 1661, 1663, and 1732. And, within New York state boundaries itself, an added 38 quakes strong enough to do damage have been recorded, with most occurring in upstate New York.

In 1857, walls vibrated, bells rang, and objects were thrown about in Buffalo from an earthquake that would be equivalent to almost a 6 on the Richter scale. In 1929, Attica, which is approximately midway between Buffalo and Rochester, was struck by an even larger earthquake, one that would be equivalent to almost a 7 on the Richter scale. According to a government report, extensive damage occurred in the Attica area: "Chimneys were thrown down, plaster was cracked or thrown down, and other building walls were noticeably damaged. Many cemetery monuments fell or were twisted. Dishes fell from shelves, pictures and mirrors fell from walls, and clocks stopped." There was also damage at Batavia. Even a wall in Sayre, Pennsylvania, over 100 miles away, cracked.

In 1944, a slightly larger quake centered just north of Massena affected an area of 450-thousand kilometers in the U.S. This quake caused more than 2 million dollars of damage in Massena and in Cornwall, Ontario to the north. For example, in Massena, masonry (90 percent of the chimneys), plumbing, and house foundations were destroyed. Many structures were unsafe for occupancy until repaired. More recently, on January 1, 1966, a 4.7 magnitude quake caused slight damage to chimneys and walls at Attica and Varysburg.

Charleston. The broad flat coastal plain of South Carolina doesn't seem to be a likely place for earthquakes. There are no known major fault systems. Nevertheless, Charleston is considered threatened because of its history of rare but devastating quakes, the mechanism of which is not known.

Since the late 1800s, Charleston has recorded 12 earthquakes strong enough to cause noticeable damage. The most recent of which was a moderate quake on November 22, 1974.

The earthquake which struck on August 31, 1886, was one of the most severe in U.S. history. At approximately 9:30 P.M., it started with a tremor, then the sound grew into a roar and finally the tremor became a rapid quiver. The first shock lasted approximately 40 seconds and a strong aftershock occurred 8 minutes later. Six additional shocks followed during the next 24 hours. Damage to the city's buildings was extensive and many buildings were totally destroyed. Railroad tracks were bent into contorted shapes and all communications were disrupted. An estimated 23 million dollars in damage was done. The cities of Columbus, South Carolina, Augusta, Georgia, and Savannah, Georgia, also experienced damage. According to a government report, "the total area affected by this earthquake covered more than 5 million square kilometers, and included such distant points as New York City, Boston, and Milwaukee in the U.S.; and Havana, Cuba and Bermuda. All or parts of thirty states and Ontario, Canada felt the principle quake." This 1886 quake would have measured about 8 on the Richter scale. Since "the causes of the 1886 Charleston shock remain an enigma" geologists wonder if "even larger events are possible" within the next few decades.

[illegible]

[illegible] Charles [illegible]

[illegible] Raymond [illegible]

[illegible] Boston [illegible] Massachusetts [illegible]

Hudson [illegible] Bermuda. All [illegible]

[illegible] few decades.

APPENDIX B:

How to Survive an Earthquake

—Adapted from *Plain Truth* (9/76) and the National Oceanic and Atmospheric Administration

---

Before an Earthquake:

1. Support building codes.
2. Encourage earthquake drills and training.
3. At home fasten shelves to walls. Remove heavy objects from upper shelves. Move beds away from windows. Have at least 2 fire extinguishers on the wall. Bolt down water heaters and other gas appliances.
4. Teach all members of your family how to turn off electricity, gas, and water at primary control points.
5. Maintain an up-to-date first aid kit. Take basic first-aid instruction.
6. Keep a flashlight with fresh batteries and a battery powered radio in the home.
7. Store extra food, water, blankets, soaps, and disinfectants in the home.
8. Work out plans of action among family members about what to do if an earthquake struck while family members were at school, work, home, or on a trip.
9. Avoid purchasing homes near known fault zones, slide areas, and below dams.
10. In new home construction or additions, try to use less glass and favor wood over brick.
11. Consider earthquake insurance and carefully read your present policy in regard to fires triggered by earthquakes.

During the Shaking:

1. Don't panic. Remain calm. Think through the consequences of any actions.

2. If indoors, stay indoors. Stay away from windows, heavy furniture that might slide or topple over, overhead fixtures, and outer walls. Stay near the center of the building and take cover under a heavy table, desk or a strong doorway. Only consider running outside if you are in a heavy, poorly constructed old building.

If in a high-rise building, do not run for exits that may become jammed or use the elevator which may fail.

3. If outside, stay in the open away from buildings and utility wires.

4. If in a moving car, stop away from utility lines and bridges and stay inside.

5. Do not use candles, matches, or other open flames.

After the Shaking:

1. Check for injuries and supply first aid. Wear shoes to avoid foot injuries from glass and debris.

2. Check for fires and fire hazards.

3. Check utilities and shut off all damaged water, electrical, and gas lines at the primary control points. Do not use or operate anything that could ignite gas until after you have made doubly sure that there are no leaks. Make sure sewage lines are intact before using toilets. Do not touch downed power lines. If your home has a chimney, inspect its entire length for cracks and damage.

4. Be careful of water and food that may have been sprayed with broken glass. If the power is off, eat the foods in your refrigerator which could spoil first. If the water is off, emergency water can be obtained from water heaters, and toilet tanks and canned foods and drinks.

5. Turn on radio or TV for emergency bulletins.

6. Do not use the telephone except for genuine emergency calls.

7. Adhere to prearranged plans about contacting and meeting family members.

8. Keep streets clear for emergency vehicles. Do not travel unless absolutely necessary.

9. Be prepared for additional aftershocks. Stay out of heavily damaged buildings. While aftershocks are usually less severe than the first shock, they can cause the collapse of structures already weakened by the first shock.

10. In certain areas be on the alert for tidal waves, landslides, or dam failure.

APPENDIX C:

Thought Forms and Biorelativity

This appendix gives a rough idea of what thought forms might do over a period of time in the environment of those producing them. Let us take a brief look at some research I performed as part of my graduate studies. In this work I was concerned with the importance of rainfall in the lives of people living with a primitive technology and possessing no well-drilling capability. I was seeking a possible relationship between their behavior and the pattern of rainfall in their environment. A scientifically air-tight argument was not necessary at that stage. Rather, I simply wanted to indicate how a broader conception of human nature, one including biorelativity, could lead to testable hypotheses. I was also looking for a theory that archaeology was better equipped to measure or test. Thought forms were taken to be electromagnetic vibrations and for the purposes of the study were grossly classified as either "harmonious" or "disharmonious" with the environment. I was looking for cases where a harmonious society got adequate precipitation and a good environment while a disharmonious society didn't. Then I would compare these findings with data which might contradict the hypothesis.

First I attempted to indicate how the physical process might operate. In December, 1971, I performed an experiment similar to those described as "the power of prayer on plants," except that the object of my investigation was water, not plants.

In a large empty room, where temperature was maintained fairly constant, four positions were established. At each position a graduated cylinder containing water was placed and the cylinder was shielded by an inverted beaker. One cylinder received prayer or positive thought from friends as they went about their normal daily activities; another was continuously exposed to classical music; another to construction, traffic, and cafeteria sounds. The fourth cylinder was not exposed to anything and served as a control. The four cylinders were periodically rotated to compensate for the factor of location. Evaporation was observed within each of the cylinders for a period of seventy-two hours. Observation showed that the cylinders exposed to the simulated "harmonious" activity—prayer and classical music—consistently evaporated more quickly than the control cylinder (prayer = +4 percent and classical music = +11 percent). The cylinder exposed to the simulated "disharmonious" activity, traffic, and other sounds evaporated slower than the control (approximately –14 percent). Despite the lack of control for various possible misleading factors, I found the results interesting and so another simple experiment was planned.

In May, 1972, this second experiment was conducted outdoors at the University of Arizona's Experimental Ranch in Oracle, Arizona. In a relatively level homogeneous field, stations were set up 100 feet apart in a checkerboard fashion. Two graduated cylinders containing water were exposed to classical music, two to prayer from friends as they went about their normal daily activities, and two cylinders served as controls. Due to mechanical problems it was not possible to expose two cylinders to simulated disharmonious activity. All cylinders were attached to metal rods set at equal heights above the foliage level and to which guages were also attached.

"Disharmonious" activity was operationally defined as the activity occurring in the presence of living plants that will reduce their growth rate below that of a control group living under the same conditions but not exposed to this activity. Thus, ten beans were planted in a pot placed at each station. All the beans were given equal amounts of water the first seven days. The experiment was run for approximately ten days. Observations showed that water in cylinders exposed to the simulated "harmonious" activity consistently evaporated at a higher rate than at the control locations, where both controls experienced the same rate of evaporation. The "prayer" cylinders evaporated approximately 4 and 14 percent faster than the controls, whereas the "classical music" cylinders evaporated approximately 4 and 21 percent faster than the controls. Plant growth was reflected by the percentage of beans planted at each location that successfully sprouted and by the cumulative height of all sprouted plants at each location.

Under the operational definition used, differences in plant growth confirmed the more "harmonious" activity at each of the prayer and classical-music locations. For example, comparing the minimum growth for each of the prayer and classical locations with the maximum growth for the control stations gave prayer an approximate 100 percent advantage in terms of sprouting and gave classical music an approximate 55 percent advantage in terms of cumulative growth. The same comparison gave classical music an approximate 50 percent advantage in terms of sprouting and an approximate 25 percent advantage in terms of cumulative growth. No rain fell during the test period so no information was gathered on suspected microrainfall differences. "Thought" was not used to reduce evaporation, but from the directional results of thought in the prayer on plant experiments I should imagine that thought could also reduce evaporation. Depending on where one lives, either more or less water evaporation and more or less rainfall would be desired. I suspect "harmonious" thoughts would effect the proper combination for the environment involved. This was a crude pilot experiment but the results suggest that something paranormal may be occurring. A much larger "checkerboard" experimental framework is planned during the rainy season of an appropriate test area.

These crude test results gain more credibility in the light of the work of several Russian government researchers. Dr. R. Tversko has worked on the dissipation of a supercooled fog in an acoustical field. As a result of ultrahigh-frequency sound vibration, a sinusoidal standing wave with points of concentration and dispersal was set up in a cloud chamber. Unconfirmed sources indicate that the Russians have succeeded in dissipating fog over airports with their ultrasonic technique. Possibly sinusoidal vibrational waves can exist within the atmospheric column above an environment, and these waves can be modified over time by human activity via resonance, evaporation, and pressure effects.

A correlation between amount of rainfall and harmonious or disharmonious activity was indicated by the archaeological sequence at three key areas in the Southwestern prehistoric record. These three sites were in the same general region: in southwest Colorado and northwest New Mexico. The first site, the Navajo Reservoir district, was occupied since the time of Christ. Subsistence was based on floodplain agriculture. Over time, the population and the number of sedentary villages steadily increased. Significant material culture changes were found in the Piedra Phase (850–950 A.D.), which included stockades, great kivas, and the introduction of religious cult paraphernalia. In addition to the stockades, evidence of warfare includes the frequency of incinerated houses and groups of human skeletons that had been cannibalized or found incinerated on pit-house floors. Population then began to concentrate more in village sites, and site distribution continuously shifted upstream due to headward entrenchment of the floodplain. This entrenchment is thought to be due to a shift to a summer-dominant precipitation pattern, as interpreted from pollen evidence, or due to an increase in winter rain, as interpreted from local paleo-microvertebrate evidence. In either case, we see that a change in rainfall is a key occurrence. This change in rainfall caused progressive desiccation of the floodplain by channel cutting, which destroyed the available agricultural fields and resulted in the total depopulation of the area by 1050.

The second site area, Chaco Canyon, was occupied since 500 A.D. by people who also lived in sedentary villages and practiced agriculture. Population and the number of villages steadily increased until by 900 activity centered within the canyon, with the development of large, multistoried town sites. Major buildings and growth in Chaco occurred from 1000 to 1100, encompassing the period of abandonment of the Navajo Reservoir district (1000–1050). By 1000, three distinct types of contemporaneous habitation sites developed: towns, villages, and McElmo pueblos. These habitation differences included location, size, type of architecture, kiva style, and luxury goods. Within a twenty-mile stretch there were thirteen towns, nine of which were within several miles of each other, all characterized by apartment-like structures that range in size from three to five stories high with 100 to 800 rooms each. There was even urban renewal of these units. On the other hand, there were almost one hundred of the relatively plain village sites. In short, it seems that there were the "haves" and the "have nots," and they were forced to live close together on the narrow strip of fertile soil along the stream in the canyon bottom. Pathological stress is indicated by social differences and the high population density. By 1130 the canyon was suddenly abandoned, and by 1150 few people remained in the canyon. During this period of abandonment, there was a drastic decrease in precipitation, as shown by arroyo cutting, tree rings, and pollen data.

The third site area, Mesa Verde, was occupied since the time of Christ and again the people lived in villages and practiced agriculture. During Pueblo II times (approximately 900–1100 A.D.), in which the Navajo Reservoir district was abandoned and in which Chaco Canyon was built up to its maximum, the Mesa Verde area was characterized by small, widely dispersed villages. In Pueblo III times (approximately 1100–1300), which encompasses the abandonment of Chaco Canyon, Mesa Verde underwent a major building period. Large multistoried pueblos were built, as well as new kinds of structures, including one platform mound of Mexican ceremonial style. Population and village size steadily increased but by 1200 villages moved from the mesa top into the numerous caves just below the mesa top in the canyon walls. Cliff dwellings of multistoried pueblos with up to two hundred rooms were built. This population shift seems to be for security, as the cliff dwellings are easily defended. Evidence of aggressive behavior also comes from the appearance of circular watch towers in the early part of this period when occupation was still on the mesa top; population density and class differences may have produced added stress. Major building occurred between 1230 and 1260, but in 1273 a prolonged drought set in until 1299, as evidenced from local tree-ring and pollen data. Since abandonment seems to have occurred at the time of the drought, it is considered the chief cause for desertion of the Mesa Verde by these agricultural Indians.

In summary, we can see that within the same general region, three localized areas at three different times show clear evidence indicating disharmonious activities: at the Navajo Reservoir district via warfare; at Chaco Canyon via population density and social stress; and at Mesa Verde via warfare, population density, and social stress. Each area then shows isolated and local environmental failure involving a change in precipitation, which in turn leads to abandonment. Similar patterns are found at the sites of Tsegi Canyon in northeast Arizona (1300 A.D.), Casa Grande in central Arizona (1400), Casas Grandes in northern Mexico (1340), Pajarito Plateau and Chama Valley in north-central New Mexico (1560–85), and at Grasshopper in east-central Arizona (1340–50). (Approximate abandonment dates have been shown in parentheses with these previous sites.) Of the many explanations offered for the abandonment of these sites, most involve at least one of the various aspects of "disharmonious" activity or environmental change incorporated into the good society–good environment–thought-form hypothesis. The Hopi Indians of north-central Arizona can be viewed as a control group. Their ethical teachings talk specifically about not causing bad vibrations and upsetting their fellow men. During the times the various areas mentioned collapsed, there was no cultural decline in the Hopi mesa area. Rather, their towns continued to grow. Through Spanish, Mexican, and U.S. sovereignty, the Hopis have kept to themselves to a fair degree and have maintained their political, social, and religious forms to this day.

This thought-form–rainfall hypothesis suggests one possible type of interaction of a set of variables we need to learn more about. Of course, rainfall or meteorological change can be caused by many environmental aspects other than the postulated thought forms. Furthermore, these other environmental aspects could override or mask the component of change from thought forms.

Note: In 1963, when a lack of snowfall threatened the financially critical Christmas ski period in Colorado, the Cloud Clan from the Ute Indian Reservation in the southwest corner of the state was called in to perform a ceremonial dance. The ski area received two feet of snow in the next three days.

In 1976 the male chorus of the San Francisco Opera Company performed a rain dance like that used by the Hopi Indians of Arizona. It rained the next day.

Bibliography

CHAPTER 1

Abrahamsen, Aron, and Abrahamsen, Doris. *Readings on Earthchanges.* Applegate, Ore., Aron and Doris Abrahamsen, 1973.

——. *Winter Newsletter.* Medford, Ore., Midoris Land, Inc., 1973.

Brown, Charles. "Quakes, Volcanoes, Killer Floods... Is Our Planet Cracking Up?" *The Star,* September 4, 1976.

Earthchanges. Professional Study Series. Virginia Beach, Va., ARE Press, 1959.

"Forecast: Earthquake." *Time,* September 1, 1975.

Goodman, Jeffrey. *Psychic Archaeology: Time Machine to the Past.* New York, Berkley/Putnam, 1977.

——. "Psychic Archaeology: Methodology and Empirical Evidence from Flagstaff, Ariz." Paper presented at the 73rd Annual Meeting of the American Anthropological Association. Mexico City, November, 1974.

Gribbin, J. R., and Plagemann, S. H. *The Jupiter Effect.* New York, Walker, 1974.

Harrison, Lee. "Psychic Predicts Exact Date and Place of Massive Killer Earthquake in New Guinea." *National Enquirer,* August 31, 1976.

James, Paul. *California Superquake.* Hicksville, N.Y., Exposition, 1974.

Lamsa, G. M. *The Holy Bible from Ancient Eastern Manuscripts.* Philadelphia, A. J. Holman, 1933.

Lindsay, H. *The Late Great Planet Earth.* Grand Rapids, Mich.; Zondervan, 1970.

Madigan, J. *World Prophecy.* Los Angeles, Mei Ling Publications, 1964.

"Quakes Hit 6-Year Record." *New Haven Register,* August 19, 1976.

Robb, Stewart. *Prophecies on World Events by Nostradamus.* New York, Liveright, 1961.

Rodenaugh, Dale. "Psychic Sees Quakes Rearranging State's Map." *San Jose Mercury,* September 27, 1976.

Scholz, C. H., Sykes, L. R., and Aggarwal, Y. P. "Earthquake Prediction: A Physical Basis." *Science*, 181, August 31, 1973, 803-809.

Taub, Miriam. "The Earthquake Lady." *Herald News,* May 6, 1977.

Tomaschek, R. "Great Earthquakes and the Astronomical Positions of Uranus." *Nature,* July 18, 1959, 177-178.

Webre, A. L., and Liss, P. H. *The Age of Cataclysm.* New York, Berkley/Putnam, 1974.

Who's Who in Science from Antiquity to the Present. Chicago, Marquis, 1968.

"The World's Climate: Unpredictable." *Time,* August 9, 1976.

"World's Quakes on Rise." *Tucson Daily Citizen,* August 20, 1974.

Ziegler, Mel. "The Earthquake Lady—Surprisingly Accurate Forecasts." *San Francisco Chronicle,* August 12, 1974.

CHAPTER 2

Abrahamsen, Aron, and Abrahamsen, Doris. *Readings on Earthchanges.* Applegate, Ore., Aron and Doris Abrahamsen, 1973.

Earthchanges. Professional Study Series. Virginia Beach, Va., ARE Press, 1959.

Elkins, Ray. "Trance State Readings." *Rays of Philosophy.* Globe, Ariz., Ray Elkins, June and November, 1976, February and March, 1977.

James, Paul. *California Superquake.* Hicksville, NY., Exposition, 1974.

Madigan, J. *World Prophecy.* Los Angeles, Mei Ling Publications, 1964.

Robb, Stewart. *Prophecies on World Events by Nostradamus.* New York, Liveright, 1961.

Webre, A. L., and Liss, P. H. *The Age of Cataclysm.* New York, Berkley/Putnam, 1974.

CHAPTER 3

Abrahamsen, Aron, and Abrahamsen, Doris. *Readings on Earthchanges.* Applegate, Ore., Aron and Doris Abrahamsen, 1973.

Canby, Thomas. "Can We Predict Quakes?" *National Geographic*, 149, June, 1976, 830-835.

Canby, Thomas, and Blair, J. P. "California's San Andreas Fault." *National Geographic*, 143, January, 1973, 38-53.

Earthchanges. Professional Study Series. Virginia Beach, Va., ARE Press, 1959.

Gilluly, J., Waters, A. C., and Woodford, A. D. *Principles of Geology*, 2nd ed. San Francisco, Freeman, 1960.

Golenpaul, Dan, ed. *Information Please Almanac.* New York, Information Please Almanac Co., 1977.

Goodman, Jeffrey. *Psychic Archaeology: Time Machine to the Past.* New York, Berkley/Putnam, 1977.

Gutenberg, B., and Richter, C. F. *Seismicity of the Earth.* Special Papers #34, New York, Geological Society of America, 1941.

Hand, S. *Geotimes,* 17:12, December, 1972.

Hopkins, D. M. *The Bering Bridge.* Palo Alto, Calif., Stanford University Press, 1967.

Ishi, I. "On the Submerged Forest of Uozer and the Subsidence of the Ground." *Journal Geography (Tokyo),* 1955, 33-43.

James, Paul. *California Superquake.* Hicksville, N.Y., Exposition, 1974.

Kisslinger, Carl. *Geotimes,* 18:30, January, 1973.

Leet, L. D., and Judson, S. *Physical Geology.* Englewood Cliffs, N.J., Prentice-Hall, 1959.

Lobeck. A. K. *Geomorphology.* New York, McGraw-Hill, 1939.

Longwell, C. R., Knopf, A., and Flint, R. F. *Physical Geology.* New York, Wiley, 1939.

Mathews, S. W. "This Changing Earth." *National Geographic,* 143, January, 1973, 1-37.

McGill, Angus. "After a Long Nap Vesuvius Awakens." *Seattle Post Intelligencer,* September 28, 1969.

Miyabe, N. "Vertical Earth Movements in Japan Deduced from Results of Relevelings." *Pacific Science Congress 7th New Zealand,* Pr 2:262-263, 1953.
National Research Council. *The Great Alaska Earthquake of 1964.* Washington, National Academy of Sciences, 1970.
Richter, C. F. "Earthquake Disasters—An International Problem." International Meeting on Earthquakes, San Francisco, NATO Committee on the Challenges of Modern Society. May, 1971.
——. *Elementary Seismology.* San Francisco, Freeman, 1968.
——. "Seismic Regionalization." *Bulletin of the Seismological Society of America,* 49, 1959, 2:123-162.
Stearn, Jess. *Edgar Cayce, The Sleeping Prophet.* Garden City, N.Y., Doubleday, 1967.
Stokes, W. L. *Essentials of Earth History—An Introduction to Historical Geology.* Englewood Cliffs, N.J., Prentice-Hall, 1960.
"U.S. Splitting Apart Says U Geophysicist." *University of Utah Review,* January, 1968.
Weber, A. L., and Liss, P. H. *The Age of Cataclysm.* New York, Berkley/Putnam, 1974.

CHAPTER 4

"African Plate an Island of Instability?" *Science News,* June 18, 1973, 393.
Ager, D. V. *The Nature of the Stratigraphical Record.* 1973.
Barbour, John. "Gentle Eastern Landscapes Hide Faults." *Arizona Daily Star,* October 24, 1976.
Dana, J. D. *Manual of Geology,* 4th ed. New York, 1894.
Evans, A. *The Palace of Minos at Knossos,* 4 vols. London, 1921–1928.
"Extra Continent May Have Existed." *Science News,* June 18, 1977, 389.
Farrand, W. R. "Frozen Mammoths and Modern Geology." *Science,* 33, March 1961, 729-735.
Gilluly, J., Waters, A. C., and Woodford, A. D. *Principles of Geology,* 2nd ed., San Francisco, Freeman, 1960.
Heim, A., and Gausser, A. *The Throne of the Gods, An Account of the First Swiss Expedition to the Himalayas,* 1939.

Hibben, F. C. *The Lost Americans.* New York, Thomas Y. Crowell, 1961.

——. "Evidence of Early Man in Alaska." *American Antiquity,* 8, 1943, 256.

Kelley, David H. "Culture Diffusion in Asia and America," in H. A. Moran and D. H. Kelley (eds.), *The Alphabet and the Ancient Calendar Signs.* Palo Alto, Calif., Daily Press, 1969.

Martinatos, S. "The Volcanic Destruction of Minoan Crete." *Antiquity,* 13, 1939, 425.

Mathews, S. W. "What's Happening to Our Climate?" *National Geographic,* 150, November, 1976, 575-615.

——. "This Changing Earth." *National Geographic,* 143, January, 1973, 1-37.

Meggers, Betty J. "The Transpacific Origin of Mesoamerican Civilization." *American Anthropologist,* 77, March, 1975, 1-28.

Ransom, C. J. *The Age of Velikovsky.* Glassboro, N.J., Kronos Press, 1976.

Stokes, W. L. *Essentials of Earth History—An Introduction to Historical Geology.* Englewood Cliffs, N.J., Prentice-Hall, 1960.

Sullivan, W. "Forecasting Disaster." *New York Times Magazine,* April 25, 1976.

——. *Continents in Motion—The New Earth Debate.* New York, McGraw-Hill, 1974.

de Terra, H., and Paterson, T. T. *Studies on the Ice Age in India and Associated Human Cultures,* 1939.

Thompson, W. I. *At the Edge of History.* New York, Harper & Row, 1971.

Velikovsky, Immanuel. *Earth in Upheaval.* Garden City, N.Y., Doubleday, 1955.

——. *Worlds in Collision.* Garden City, N.Y., Doubleday, 1950.

Walworth, R. F. *Subdue the Earth.* New York, Delacorte, 1977.

West, R. G. *Pleistocene Geology and Biology.* New York, Wiley, 1968.

CHAPTER 5

Canby, T. "Can We Predict Quakes?" *National Geographic*, 149, June, 1976, 830-835.

"Forecast: Earthquake." *Time,* September 1, 1975.

Hempel, C. G. *Aspects of Scientific Explanation.* New York, Free Press, 1965.

"California Quake Watch Need Cited." *Tucson Daily Citizen,* May 4, 1976.

"Lower Colorado Game Habitat to Be Lost." *Arizona Daily Star,* July 3, 1977.

"Old Faithful Quake Warnings." *Time,* October 18, 1971.

Richards, J., et al. *Modern University Physics.* Reading, Mass., Addison-Wesley, 1960.

Scholz, C. H., Sykes, L. R., and Aggarwal, Y. P. "Earthquake Prediction: A Physical Basis." *Science,* 181, August 31, 1973, 803-809.

Sullivan, W. "Forecasting Disaster." *New York Times Magazine,* April 25, 1976.

Vogt, F. R. "Evidence for Global Synchronism in Mantle Plume Convection and Possible Significance for Geology." *Nature,* 240, December 8, 1972, 338-342.

"Well Water Level Predicts Quakes." *Tucson Daily Citizen,* June 24, 1976.

CHAPTER 6

"Another Ice Age?" *Time,* June 24, 1974.

"Brr, Yes, It's a Record Cold U.S. Winter." *Science News,* February 26, 1977, 139.

Bryson, Reid. *Climates of Hunger.* Madison, University of Wisconsin Press, 1977.

"Drought Hurting Brazil." *Tucson Daily Citizen,* November 22, 1976.

Hays, J. D., Imbrie, J., and Shackelton, N. J. "Variations in the Earth's Orbit: Pacemaker of the Ages." *Science,* 194, December 10, 1976, 1121-1132.

"Ice Ages Attributed to Orbit Changes." *Science News,* 110, 1976, 356.
Impact Team. *The Coming of the New Ice Age.* New York, Ballantine, 1977.
"Is the Climate Trying to Tell Us Something?" *Arizona Daily Star,* September 15, 1974.
"Just Trying to Survive." *Time,* July 11, 1977.
Lamb, H. H. *The Changing Climate.* London, Methuen, 1966.
"Major Famines Predicted—Big Climatic Changes Forecast." *Tucson Daily Citizen,* April 1, 1976.
"Man Altering Climate?" *Tucson Daily Citizen,* December 29, 1975.
Mathews, S. W. "What's Happening to Our Climate?" *National Geographic,* 150, November, 1976, 576-615.
Mitchell, J. M. "Theoretical Paleoclimatology," in *The Quaternary of the United States.* H. E. Wright and David Frey (eds.), Princeton, N.J., Princeton University Press, 1965.
"Scientists Foresee Cooler Climates." *Los Angeles Times,* December 1, 1976.
Sellers, W. D. *Physical Climatology.* Chicago, University of Chicago Press, 1969.
Stokes, W. L. *Essentials of Earth History—An Introduction to Historical Geology.* Englewood Cliffs, N.J., Prentice-Hall, 1960.
"World's Climate Unpredictable." *Time,* August 9, 1976.

CHAPTER 7

Barbetti, M. McElhinny. "Evidence of a Geomagnetic Excursion 30,000 Years BP." *Nature,* 239, December 27, 1971, 327-330.
Dunbar, C. O. *Historical Geology.* New York, Wiley, 1949.
Durrant, S. A., and Khan, H. A. "Ivory Coast Tektites, Fission Track Ages and Geomagnetic Reversals." *Nature,* 232, 1971, 320.
Glass, B., and Heezen, B. C. "Tektites and Geomagnetic Reversals." *Scientific American,* 217, July 1967, 32-38.
Gold, Thomas. "Instability of the Earth's Axis of Rotation." *Nature,* 175, March 26, 1975, 526.

Hapgood, Charles H. *Earth's Shifting Crust.* New York,, Chilton, 1958.

——. *The Path of the Pole.* Philadelphia, Chilton, 1970.

Hays, James D. "Faunal Extinction and Reversals of the Earth's Magnetic Field." *Geological Society of America Bulletin,* 82, September, 1971, 2433-2447.

Hays, James D., and Updyke, N. "Antarctic Radiolaria, Magnetic Reversals and Climate Change." *Science,* 158, November 24, 1967, 1001-1011.

Kawai, N., et al. "Oscillating Geomagnetic Fields with Recurring Reversals Discovered for Lake Biwa." *Proceedings of the Japanese Academy*, 48, March, 1972, 186-190.

Kennet, J. P., and Watkins, N. D. "Geomagnetic Polarity Changes, Volcanic Maxima and Faunal Extinctions." *Nature,* 227, 1970, 930.

Knopoff, L., and Leeds, A. "Lithospheric Momenta and the Deceleration of the Earth." *Nature,* 237, May 12, 1972, 93.

Lamsa, G. M. *The Holy Bible from Ancient Eastern Manuscripts.* Philadelphia, A. J. Holman, 1957.

Mansinha, L., and Smylie, D. E. "The Rotation of the Earth." *Scientific American,* December, 1971, 80-89.

——. "Earthquakes and the Earth's Wobble." *Science,* September 13, 1967, 1127.

Morner, N. A., et al. "Late Weichselian Paleomagnetic Reversal." *Nature,* 234, December 27, 1971, 441.

Munk, W., and Revelle, R. "Sea Level and the Rotation of the Earth." *American Journal of Science,* 250, November, 1952, 829-833.

O'Neill, J. J. "Atomic Energy Changing Globe, Held Able to Erupt at Any Time." *New York Herald Tribune,* August 11, 1946.

Opdyke, W. D., et al. "Paleomagnetic Study of Antarctic Deep-Sea Cores." *Science,* 154, 1966, 349.

Ransom, C. J. *The Age of Velikovsky.* Glassboro, N.J., Kronos Press, 1976.

Stokes, W. L. *Essentials of Earth History—An Introduction to Historical Geology.* Englewood Cliffs, N.J., Prentice-Hall, 1960.

Sullivan, W. *Continents in Motion—The New Earth Debate.* New York, McGraw-Hill, 1974.

Thorpe, S.A. "A New Magnetic Reversal at 12,500 Years?" *Nature,* 234, December 24, 1971, 441.
Valentine, Tom. *The Life and Death of Planet Earth.* New York, Pinnacle, 1977.
Velikovsky, Immanuel. *Earth in Upheaval.* Garden City, N.Y., Doubleday, 1955.
——. *Worlds in Collision.* Garden City, N.Y., Doubleday, 1950.
Waddington, C. J. "Paleomagnetic Field Reversals and Cosmic Radiation." *Science,* 158, November 17, 1967, 913.
Waters, Frank. *The Book of the Hopi.* New York, Ballantine, 1963.
Watkins, N. D., and Goodell, H. G. "Geomagnetic Polarity Change and Faunal Extinction in the Southern Ocean." *Science,* 156, May 26, 1967, 1083-1087.

CHAPTER 8

Blizard, J. B. "Predictions of Solar Flares Months in Advance." *Astronomical Journal,* 70, 1965, 667.
——. "Long Range Solar Flare Prediction." *Astronomical Journal,* 73, 1968.
Bureau, R. R., and Crane, L. B. "Sunspots and Planetary Orbits." *Nature,* 228, December 5, 1970, 984.
Dewey, E. R. "A Key to Sunspot-Planetary Relationship." *Cycles,* October, 1968.
Dobrin, M. *Introduction to Geophysical Prospecting,* 2nd ed. New York, McGraw-Hill, 1960.
Earthchanges, Past, Present, Future. Professional Study Series, No. 1. Virginia Beach, Va., ARE Press, 1959.
"Earthquakes and the Moon." *Science News,* January 22, 1977, 51.
Gribbin, J. "Relation of Sunspot and Earthquake Activity." *Science,* 173, 1972, 558.
Gribbin, J., and Plagemann, S. H. *The Jupiter Effect.* New York, Walker, 1974.
Gutenberg, B., and Richter, C. R. *Seismicity of the Earth.* Special Papers #34, Geological Society of America, New York, 1941.

Hunter, R. N. "Is There a Connection Between Astrology and Earthquakes?" *Earthquake Information Bulletin,* May-June 1971.

"Jupiter's 'Vanishing' Magnetic Fields Baffling Scientists." *Arizona Daily Star,* December 2, 1973.

Mansinha, L., and Smylie, D. E. "Earthquakes and the Earth's Wobble." *Science,* September 13, 1968, 1127-1129.

——. "The Rotation of the Earth." *Scientific American,* December, 1971, 80-89.

"Mars: A Possible New Quake." *Science News,* January 29, 1977, 68.

"Mercury Unveiled." *Time,* April 8, 1974.

"Mercury's Magnetism." *Time,* March 31, 1975.

"Move over Olympus Mons—Here Comes Beta." *Science News,* May 14, 1977, 245.

Nelson, J. H. "Short-wave Radio Propagation Correlation with Planetary Positions." *RCA Review,* March 1951, 26-34.

——. "Planetary Position Effect on Short-Wave Signal Quality." *Electrical Engineering,* May 1952, 421-424.

"News Under the Sun—Jupiter's Tail." *Time,* May 3, 1976.

"Rings Around Uranus." *Time,* April 11, 1977.

Sanford, F. "Influence of Planetary Configurations Upon the Frequency of Visible Sunspots." *Smithsonian Miscellaneous Collections,* 95, 11, 1935, 15.

Shuvalov, V. "Solar Activity and Positions of the Planets." *Nauka i Zhizn,* 10, 1971, 63-64.

"Solar Flares: Link to Thunderstorms." *Science News,* June 18, 1977, 389.

Stecchini, Livio. "Notes on the Rotation of Ancient Measure to the Great Pyramid," in Peter Tomkins, *Secrets of the Great Pyramid.* New York, Harper & Row, 1971.

"Strange Rings of Uranus." *Science News,* April 16, 1977, 245.

Takahashi, K. "On the Relation Between the Solar Activity Cycle and the Solar Tidal Force Induced by the Planets." *Solar Physics,* 3, 1968, 598-602.

Tomaschek, R. "Great Earthquakes and the Astronomical Positions of Uranus." *Nature,* July 18, 1959, 177-178.

——. "Reply." *Nature,* April 23, 1960, 337-338.

U.S. Department of Commerce. "Earthquake History of the United States." Publication 41-1, *National Oceanic and*

Atmospheric Administration Environmental Data Service, 1973.
Valentine, Tom. *The Life and Death of Planet Earth.* New York, Pinnacle, 1977.
Velikovsky, Immanuel. *Worlds in Collision.* Garden City, N.Y., Doubleday, 1950.
Weaver, K. F. "Journey to Mars." *National Geographic,* 143, February, 1973, 231-263.

CHAPTER 9

Backster, C. "Evidence of Primary Perceptions in Plant Life." *International Journal of Parapsychology,* 10, 4, Winter, 1968.
Bartlett, Laile. "What Do We Really Know About Psychic Phenomena?" *Reader's Digest,* April, 1977.
Cohen, D. "Magnetoencephalography: Detection of the Brain's Electrical Activity—a Superconduction Magnetometer." *Science,* 175, 1972, 664-666.
——. "Magnetocardiography of Direct Currents: S-T and Baseline Shifts during Experimental Myocardial Infarction." *Science,* 172, 1968, 1329-1333.
——. "Magnetic Fields Around the Torso: Production by Electrical Activity of the Human Heart." *Science,* 156, 1967, 652-654.
Lamsa, G. M. *The Holy Bible from Ancient Eastern Manuscripts.* Philadelphia, A. J. Holman, 1933.
Loehr, F. *The Power of Prayer on Plants.* New York, New American Library, 1969.
Ostrander, S., and Schroeder, L., *Psychic Discoveries Behind the Iron Curtain.* Englewood Cliffs, N.J., Prentice-Hall, 1970.
Tiller, W. "A General Technical Report on A.R.E. Fact-Finding Trip to the Soviet Union." *ARE Journal,* 7, March, 1972, 68-81.
Tomkins, P., and Bird, C. *The Secret Life of Plants.* New York, Harper & Row, 1973.
Tversko, N. P. "Dissipation of a Supercooled Fog in an

Acoustical Fluid." *Tetrady Glavnoi Geofig. Obs.*, 176:51-69, Leningrad, 1965.
Waters, Frank. *The Book of the Hopi*. Ballantine Books, New York, 1963.
Watson, L. *Supernature*. Garden City, N.Y., Doubleday, 1973.
Note: Also see Bibliography for Appendix C.

CHAPTER 10

"Animals, Not Instruments, May Predict Quakes." *Washington Post,* November 25, 1976.
Canby, T. "Can We Predict Quakes?" *National Geographic*, 149, June, 1976, 830–835.
Cerminara, Gina. "Missie—The Psychic Dog." *Psychic,* October, 1973.
"China: Shock and Terror in the Night." *Time,* August 9, 1976.
James, Paul. *California Superquake*. Hicksville, N.Y., Exposition, 1974.
Morris, R. L. "The Psychobiology of Psi," in *Psychic Explorations*. Edgar Mitchell (ed.), New York, Putnam, 1976.
——. "Psi and Animal Behavior: A Survey." *Journal of American Society of Psychical Research,* 64, 1970, 242-260.
"Psychic Chimps Amaze Experts After Predicting Two Earthquakes." *The Star,* December 14, 1976.
Rhine, J. B. "The Present Outlook on the Question of Psi in Animals." *Journal of Parapsychology,* 15, 1951, 230-251.
Rhine, J. B., and Feather, S. R. "The Study of Cases of Psi-Trailing in Animals." *Journal of Parapsychology,* 26, 1962, 1-22.
"Scientist Now Believes Odd Animal Behavior Might Predict Earthquakes." *New York Daily News,* December 20, 1976.
Watson, L. *Supernature*. Garden City, N.Y., Doubleday, 1973.
Webster, Bayard. "Animals Are Tested on Quake Forecasts." *New York Times,* April 14, 1977.
Werner, Alex. "Russian Scientists Use Animals to Predict National Disasters." *Midnight,* July 14, 1969.

CHAPTER 11

Abrahamsen, Aron, and Abrahamsen, Doris. *Readings on Earthchanges.* Applegate, Ore., Aron and Doris Abrahamsen, 1973.

Dakenbring, W. F. "The Great Quake: When Will It Come?" *Plain Truth,* November 22, 1975.

Dart, J., and Chandler, R. "Is Armageddon at Hand?" *Tucson Daily Citizen,* August 14, 1976.

Furst, Jeffrey. *Edgar Cayce's Story of Jesus.* New York, Coward McCann Geoghegan, 1968.

Lamsa, G. *The Holy Bible from Ancient Eastern Manuscripts.* Philadelphia, A. J. Holman, 1933.

Lindsey, H. *The Late Great Planet Earth.* Grand Rapids, Mich., Zondervan, 1970.

Simpson, G. G. *The Meaning of Evolution.* New Haven, Conn., Yale University Press, 1966.

Valentine, Tom. *The Life and Death of Planet Earth.* New York, Pinnacle, 1977.

Waters, Frank. *The Book of the Hopi.* New York, Ballantine, 1963.

CHAPTER 12

"California Quake Watch Need Cited." *Los Angeles Times,* May 4, 1976.

"Californians Told to Be Prepared for Great Quake." *Tucson Daily Citizen,* December 10, 1976.

Canby, T. "Can We Predict Quakes?" *National Geographic,* 149, June, 1976, 830-835.

——. "California's San Andreas Fault." *National Geographic,* 143, January, 1973, 38-53.

"Chinese Make Progress in Quake Predicting." *Tucson Daily Citizen,* January 14, 1976.

"Earthquake Jitters." *Time,* March 8, 1971.

Eckert, Allan W. *The HAB Theory.* Boston, Little Brown, 1976.

"Forecast: Earthquake." *Time,* September 1, 1975.

"Forecast: Future Shock." *Time,* January 24, 1977.
"Great Depression a Great Leveler." *Arizona Daily Star,* October 27, 1974.
Keeler, Sussi. "Government Computers Will Test Accuracy of Psychics Who Predict Earthquakes." *National Enquirer,* 1976.
Mayhew, Pamela. "Is There an Earthquake in Our Future?" *Tucson Daily Citizen,* March 29, 1975.
"Measuring Disaster." *Time,* October 11, 1976.
"Predicting Earthquakes." *Tucson Daily Citizen,* August 17, 1977.
"Quake Experts Worried by South California Uplift." *Tucson Daily Citizen,* February 15, 1976.
"Quakes May Have Caused Red A-Blast." *Tucson Daily Citizen,* November 18, 1976.
"Quakes Would Fell L.A. Old Buildings." *Arizona Daily Star,* February 6, 1977.
"Social Hazards of Earthquake Prediction." *Science News,* January 8, 1977, 20.
"Southern Californians Urged to Prepare for Earthquake." *Los Angeles Times,* April 20, 1976.
Sullivan, Walter. "Forecasting Disaster." *New York Times Magazine,* April 25, 1976.
"Tangshan Quake: Portrait of a Catastrophe." *Science News,* June 18, 1977, 388.
"Tracking Quake Data and Predictions." *Science News,* December 11, 1976, 37.

APPENDIX C

Backster, C. "Evidence of Primary Perceptions in Plant Life." *International Journal of Parapsychology,* 10, 4, Winter, 1968.
Baule, G., and McFee, R. "Detection of the Magnetic Field of the Heart." *American Heart Journal,* 66, 1963, 95-96.
Boas, F. *Race, Language and Culture.* New York, Free Press, 1966.
Calhoun, J. B. "Population Density and Social Pathology." *Scientific American,* 206, 1962, 139-148.

Carey, G. "Density, Crowding, Stress and the Ghetto." *American Behavioral Scientist,* 1972, 496-509.

Changnor, S., and Huff, F. "Metromex: An Investigation of Inadvertent Weather Modification." *Bulletin of the American Meteorological Society,* 52, 1972, 958-967.

Cohen, D. "Magnetoencephalography: Detection of the Brain's Electrical Activity—a Superconduction Magnetometer." *Science,* 175, 1972, 664-666.

——. "Magnetocardiography of Direct Currents: S-T and Baseline Shifts During Experimental Myocardial Infarction." *Science,* 172, 1968, 1329-1333.

——. "Magnetoencephalography: Evidence of Magnetic Fields by Alpha-Rhythm Currents." *Science*, 161, 784–786.

——. "Magnetic Fields Around the Torso: Production of Electrical Activity by the Human Heart." *Science,* 156, 652-654.

Dean, J. "Aspects of Tsegi Phase Social Organization: A Trial Reconstruction," in Longacre, W. A. (ed.), *Reconstructing Prehistoric Pueblo Society.* Albuquerque, University of New Mexico Press, 1970.

Di Peso, "Cultural Development in New Mexico's Aboriginal, Cultural Development in Latin America: An Interpretative Review." Meggers and Evans (eds.), *Smithsonian Miscellaneous Collections*, 146, 1, 1970.

Douglass, A. E. "Dating Pueblo Bonito and Other Ruins of the Southwest." *National Geographic Society,* Bonito Series, 1, 1935.

Eddy, F. W. "Prehistory in the Navajo Reservoir District." *New Mexico, Papers in Anthropology,* 3:15, pts. 1, 2, Albuquerque, Museum of New Mexico, 1966.

Eisenberg, L. "The Human Nature of Man." *Science*, 176, 123–128.

Flannery, K. "The Cultural Evolution of Civilization." *Annual Review of Ecology and Systematics,* 3, 1972.

Fritts, Smith, and Stokes. "The Biological Model for Paleoclimatological Interpretation of Mesa Verde Tree Ring Series." *American Antiquity,* 31, 1965, 101-122.

Fritz, J., and Plog, F. "The Nature of Archaeological Explanation." *American Anthropologist,* 35, 1970, 405-412.

Galle, O., and Grove, W. "Population Density and Pathology:

What Are the Relations for Man?" *Science*, 176, 1972, 23-30.

Gladwin, W., and Haury, E. "Excavations at Snaketown, Material Culture, Gila Pueblo." *Medallion Papers,* 25, 1937.

Harris, A. "Past Climate of the Navajo Reservoir District." *American Antiquity,* 135, 1970, 374-377.

Harris, M. *The Rise of Anthropological Theory.* New York, Thomas Y. Crowell, 1968.

Jett, S. C. "Pueblo Indian Migration: An Evolution of the Possible Physical and Cultural Determinants." *American Antiquity,* 29, 1964, 281-300.

Johnson, L. "Problems in Avant Garde Archaeology." *American Anthropologist,* 73, 74, 1972, 366–377.

Kuhn, T. S. *The Structure of Scientific Revolutions.* Chicago, University of Chicago Press, 1970.

Lamsa, G. M. *The Holy Bible from Ancient Eastern Manuscripts.* Philadelphia, A. J. Holman, 1933.

Lee, R. "King Bushmen." *Canadian National Museum Bulletin,* 1966.

Loehr, F. *The Power of Prayer on Plants.* New York, New American Library, 1969.

Luebben, H. "An Unusual PIII Ruin, Mesa Verde." *American Antiquity,* 26, 1957, 1121.

Mitchell, J. M. "Theoretical Paleoclimatology," in *The Quaternary of the United States.* H. E. Wright and David Frey (eds.), Princeton, N.J., Princeton University Press, 1965.

Montagu, A. *The Humanization of Man.* New York, World, 1962.

Ostrander, S., Schroeder, L., *Psychic Discoveries Behind the Iron Curtain.* Englewood Cliffs, N.J., Prentice-Hall, 1970.

Pressman, A. S. "Electromagnetic Fields and Living Nature." *Science,* Moscow, 1968.

Richards, J., et al. *Modern University Physics.* Reading, Mass., Addison-Wesley, 1960.

Rohn, P. *Mug House.* National Park Service—Archaeological Research Service #7D, 1971.

Shoenwetter, J., and Dittert, A. "An Ecological Interpretation of Anasazi Settlement Pattern," in Meggers (ed.), *Anthropological Archeology in the Americas.* Washington, 1968.

Simpson, G. G. *The Meaning of Evolution.* New Haven, Conn., Yale Univ. Press, 1966.

Streuver, S. *Prehistoric Agriculture.* New York, Natural History Press, 1971.

Tiller, W. "A General Technical Report on A.R.E. Fact Finding Trip to the Soviet Union," *ARE Journal*, 11, 68–81. Virginia Beach, Va., ARE Press, 1972.

Tversko, N. P. "Dissipation of Supercooled Fog in an Acoustical Fluid." *Tetrady Glavnoi Geofig. Obs.*, 176, 1965, 51-69.

Vivian, R. *An Inquiry into Prehistoric Social Origins in Chaco Canyon.* Albuquerque, U. of New Mexico Press, 1970.

Vivian, R. G., and Mathews, T. *Kin Kletso, A Pueblo III Community in Chaco Canyon.* Globe, Ariz., Southwestern Monument Association, Technical Series, 6, pt. 1, 1965.

Waters, Frank. *The Book of the Hopi.* New York, Ballantine Books, 1963.

Wendorf, F., and Reed, E. "An Alternative Reconstruction of Rio Grande Prehistory." *El Palacio,* 62, 1955, 131-173.

White, J. "Plants, Polygraphs and Paraphysics." *Psychic,* 4, 2, 1972, 13-17.

Index